R. J. Amman

Build It Underground

A Guide for the
Self-Builder
& Building Professional

Build It Underground

A Guide for the Self-Builder & Building Professional

Introduction by Robert L. Roy

David Carter

Sterling Publishing Co., Inc. New York

Photo Credits
The author and publisher wish to thank the following for the use of their photographs:
BRW Architects, pages 191, 192, 193
Julie Carter, pages 47, 66, 178
Leo A. Daly Co, page 188
Jim and Sher Dostal, pages 140, 141, 142
Dick Kennedy, pages 186, 187
The Minnesota Underground Space Association, page 182

Library of Congress Cataloging in Publication Data

Carter, David.
   Build it underground.

   Includes index.
   1. Underground construction. I. Title.
TA712.C257   690   81-85021
ISBN 0-8069-5454-X   AACR2
ISBN 0-8069-5455-8 (lib. bdg.)
ISBN 0-8069-7582-2 (pbk.)

Copyright © 1982 by David Carter
Published by Sterling Publishing Co., Inc.
Two Park Avenue, New York, N.Y. 10016
Distributed in Australia by Oak Tree Press Co., Ltd.
P.O. Box K514 Haymarket, Sydney 2000, N.S.W.
Distributed in the United Kingdom by Blandford Press
Link House, West Street, Poole, Dorset BH15 1LL, England
Distributed in Canada by Oak Tree Press Ltd.
% Canadian Manda Group, 215 Lakeshore Boulevard East
Toronto, Ontario M5A 3W9
Manufactured in the United States of America
All rights reserved

Contents

Metric Conversion Chart   7

Foreword   8

Acknowledgments   8

Introduction   9

1. A Historical Perspective   11

2. Preplanning Considerations   18

3. Permits, Insurance and Financing   27

4. Purchasing Land   33

5. Basic Earth-Shelter Modes   37

6. Prairie Cocoon   45

7. Trench House   72

8. Railroad Tie House   84

9. Deep Woods A-Frame   94

10. Mobile Home Earth Shelters   101

11. Round Concrete Structures   109

12. Convertible Crescent   128

13. Starlight Earth Lodge   132

14. Cordwood Courtyard   138

15. Easy Arch System   145

16. Double Hex   149

17. Wondrous Warren   154

18. Stairway House   158

19. Geotecture: Self-Supporting TunnelledSpace   160

20. Energy-Efficient Systems   164

21. Landscaping   169

22. Spatial Perceptions and Uses   173

23. Commercial Applications   180

Glossary   197

Preplanning Checklist   204

Builder's Checklist   204

Index   205

Metric Conversion Chart

(Approximate Conversion Factors)

To change	To	Multiply by
inches	centimetres	2.540
feet	metres	.305
yards	metres	.914
miles	kilometres	1.609
square inches	square centimetres	6.451
square feet	square metres	.093
square yards	square metres	.836
cubic feet	cubic metres	.028
cubic yards	cubic metres	.765
fluid ounces	millilitres	29.573
pints	litres	.473
quarts	litres	.946
gallons	litres	3.785
ounces	grams	28.349
pounds	kilograms	.454
centimetres	inches	.394
metres	feet	3.280
metres	yards	1.094
kilometres	miles	.621
square centimetres	square inches	.155
square metres	square feet	10.764
square metres	square yards	1.196
cubic metres	cubic feet	35.315
cubic metres	cubic yards	1.308
millilitres	fluid ounces	.034
litres	pints	2.113
litres	quarts	1.057
litres	gallons	.264
grams	ounces	.035
kilograms	pounds	2.205

Foreword

I have written this book with many ideas and purposes in mind. My main purpose is to provoke a feeling that living and working underground makes sense and can be a lot of fun.

I have tried to include information of every type that will be beneficial in planning, building and enjoying underground space. Almost all of my designs are low-cost and can be self-built. The designs use either indigenous material, recycled material or material that is universally available at competitive prices.

The decision to act on this information will be up to the reader. This special experience comes in many shapes, sizes, materials, uses and values. Those feelings from childhood as we dug our forts and hideouts in the backyard and pasture were probably right after all.

Acknowledgments

No one is isolated from the thoughts, dreams and plans of others. I am privileged to have learned most of the things included in this book from many gracious people. Mike Oehler's pioneering efforts in low-cost post-and-beam underground dwellings have certainly influenced my thinking. Mike Hackelman's mastery of the wind and knowledge of wind-electrics have convinced me that this is the only way to go when building underground—or on the surface. Jack Henstridge's success with cordwood construction has given me added emphasis on using indigenous material wherever I build. David and Lydia Miller's fantastic success using rammed earth to build has inspired a number of successful underground experiments on my part.

Emil Christensen, a man who has walked and preserved the land as no other man I know, has been years ahead of his time in using the earth directly for shelter, recreation and living. My tunnelling projects were directly inspired by this gentle man. Rob Roy has given me inspiration to encourage others towards self-sufficiency in building and living. All of my friends at *Mother Earth News* magazine played no small part in my inspiration and are to be applauded. My personal friends, Warren Gilbert, Gary Everett and Jim Brown, and my parents have encouraged me in all of my projects when many others would have given up long ago. Paul Isaacson's thoughts and friendship have been helpful aids in my teaching at *Mother Earth*.

Production help on this book was every bit as important as my writing in producing a readable and visually pleasing product. Carol Witfoth not only typed the manuscript and helped correct my many errors, but also encouraged me to keep writing when I was about ready to throw in the towel. Without Jim Brown's nationally acclaimed art talents and personal help, this book would not have been produced. All of you, and you know who you are, who have smiled and helped through the years, thank you!

A special thanks to my three daughters, Julie, Leslie and Heidi, for keeping their dad going when everything looked impossible. I love you.

Introduction

The "underground" movement continues its almost geometric growth since its rebirth—or at least repopularization—during the mid-Seventies. While only a handful of earth-sheltered houses were in existence in the United States ten years ago, the number today is into the several thousands. The range of kinds and qualities of structures—and *costs* (not always synonymous with quality)—is as varied as for surface structures. For this reason, it has become necessary for those of us working in the field to focus his or her energies in specific directions within the larger whole. Unfortunately, the profit motive is all too prevalent in the work of many designers. It does my heart good, therefore, that Dave Carter's point of view throughout this book is always of particular interest to the "little guy" who wants to own his own home without being shackled by the "gold and silver fetters" of a thirty-year mortgage.

I recently wrote an article for *Earth Shelter Digest* pointing out that all too many of the better-known earth-shelters are built and sold in the $100,000-and-up class, creating the misconception that underground housing is only for the very rich. Editor Kathy Vadnais was glad to receive news that the "masses of men" are still out there, thirsting for information, *and building*. ("There are more of us than there are of them," says Jack Henstridge, champion of affordable owner-built housing.) Both Ms. Vadnais and the redoubtable Henstridge should be delighted with Dave's current contribution.

Dave's designs are imaginative, it is true, but always his imagination is based on real work which has been tried and tested in the field—or *beneath* the field—by Dave himself in some cases, and by innovators such as Mike Oehler, Paul Isaacson and Emil Christiansen in others. Dave is particularly careful to avoid the trap of being all tied up in past- or present-day technology. He has the knack of taking the appropriate systems from past and present and knitting them into a workable and affordable fabric for the self-builder.

For the building professional, the designs may help to free the thinking from the prevalent earth-shelter plans resembling motel units plugged into a hillside, and towards more harmonious and cost-effective methods using *indigenous materials*. Animal shelters built by birds, bees, badgers and beavers harmonize with their environment for precisely that reason, often so successfully that their structures are difficult to spot in nature's perfect setting. Underground housing is perhaps our best opportunity to do the same.

Probably the most valuable contribution that can be made by a writer in the building field is to report his mistakes as accurately as his successes. Too many writers miss this golden opportunity to help the next guy from falling into the same holes, no pun intended. Every builder working with innovative structures will make mistakes. I'm instantly suspicious of building books that paint too rosy a picture and are conspicuously lacking in problems. The truly conscientious researchers and experimental builders, like Dave Carter, do not place themselves above reporting problems and mistakes, and advising how the particular situation could have been handled better. Dave's building experiences at his Prairie Cocoon and elsewhere put the reader in touch with the very real on-the-spot considerations that exist with innovative construction.

I see *Build It Underground* as a mind-freeing work, steeped with plenty of nuts-and-bolts stuff not found in other earth-shelter books. As Dave cautions time and time again, the self-builder should have his individual plans checked by a competent structural engineer prior to construction. Overbuilding is unnecessarily expensive. Underbuilding is disastrous. Therefore, retaining an engineer is both economic and prudent.

Robert L. Roy
Log End Cave
West Chazy, New York

1
A Historical Perspective

Historic underground dwellings are found in nearly every part of the world. These timeless abodes continue to house their occupants in a temperate, energy-independent environment. Opal miners in Australia tunnel elaborate homes in the sandstone, complete with swimming pools. The Soviet Union, the People's Republic of China, India, Spain, Greece, Italy, Turkey, Tunisia and many other countries have large populations living underground.

Two factors usually determine the success of any given design. The main factor is the design's continued usefulness in satisfactorily sheltering its inhabitants in a given environment. The second factor is the availability and cost of building material. The Tunisian design is a good example of these factors: It is still the best design for that environment and is constructed from the earth itself. The design of Tunisian underground dwellings dates back over a thousand years and evolved out of two needs: overcoming the harsh climate and providing protection from enemies. Craters were dug and individual dwellings were tunnelled into the crater's walls. Narrow tunnels and entrances were dug to allow only one person to enter at one time. This technique made the dwellings easy to defend, and their depth below the surface made them temperate and stable. These craters varied in size from thirty feet in diameter to over one hundred feet. This design concept is still in use on a large scale in Tunisia.

A large portion of China's population lived in underground dwellings thousands of years ago. The "courtyard" shelters then used were functional and attractive. Several dwellings were constructed around the courtyard and provided a common use area for several families. This design provided for a good utilization of space while giving a feeling of openness. During the height of the Roman Empire, the Romans added elaborate hanging gardens and columns to the earth above the Tunisian homes, but otherwise left them the same.

Underground construction is man's attempt to start cooperating with nature for his survival; therefore, he needs to realize that everything is interrelated and interdependent. The North American Indians held this tenet more closely than any other group. In 1855, one of the great chiefs, Chief Seathl of the Duwamish tribe, wrote a letter to the President of the United States, Franklin Pierce, during a time when the white man was seeking to buy land that the Indian used but never claimed to own. Chief Seathl's letter is probably one of the most moving documents ever written in defense of cooperating with nature for survival, rather than trying to rule nature.

When North Americans think of their historical homes, they are most apt to think of salt box colonial homes or log cabins. Many thousands of earth homes were built on the Great Plains of the United States during the exploration of the West which were either sod-block construction or dugouts. The availability of building material and the environmental conditions dictated design.

TUNISIAN CRATER DWELLINGS

The Indians who inhabited the Great Plains lived in very efficient earth lodges before the white man ever crossed the Mississippi River. Prehistoric earth lodges have been found in large numbers throughout Nebraska, North Dakota and South Dakota. These structures varied greatly in size and shape. They all utilized a post-and-rafter system which supported a sheeting of smaller branches that was covered with earth. A thatched-grass shingle system was utilized over the sheeting cover to shed any water that seeped through the earth cover. A smoke hole in the center of the lodge accommodated a cooking fire. An entrance tunnel added storage space, while making the lodge easy to defend.

Many of the principles utilized in the construction of the early earth lodges and the pioneers' dugouts are still valid today. Rob Roy and Stu Campbell both use the post-and-beam method of construction for their homes. The use of indigenous building materials and design concepts is sound both structurally and economically. Pioneer dugouts became more sophisticated as the nineteenth century came to a close. The twentieth century saw many of these underground homes rivalling some of today's designs. The use of cellulose insulation, plastered walls, hardwood floors and the availability of glass made these dwellings dry, light and attractive. Some of the designs involved digging into a bank and using posts, lumber shoring and rafter systems to support sod roofs. Other designs relied on posts to hold log walls against earth berms. A shallow "A"-type roof was then covered with sod.

Pitch was used for waterproofing when available. A more natural and effective waterproofing was discovered and used in the early 1900's. Bentonite, a very fine clay, made the best waterproofing material, and it is still an excellent waterproofing substance today. This clay is milled into a dry powder, combined with water, and trowelled onto the structure. The structure is then backfilled before the bentonite can dry. Bentonite swells into a water barrier that keeps the larger water molecules from penetrating. It stays moist and expanded by drawing moisture from surrounding soil, and its real beauty lies in its inability to deteriorate. Modern waterproofing products sandwich powdered bentonite between two layers of cardboard. These panels are then installed on the structure, irrigated and backfilled.

A new generation of dugout pioneers developed in the Seventies due to the energy crunch. They relearned what the original pioneers could have taught them about structural systems, waterproofing, insulation and the thermal properties of the earth. My own research to locate any helpful information that would keep me from making too many mistakes while constructing my first home led me to the Nebraska State Historical Society, where I found dozens of photographs of pioneer dugouts. Personal letters and other memorabilia also consigned to the Society by descendants of these pioneer builders provided additional help. Understanding some of the problems as well as advantages of underground building and living became more clear.

Several letters written by these pioneers to their relatives indicate some of the problems they encountered. One letter described water pouring in through the wall during a heavy storm. The cause of this sudden flood was located after the storm. It seems that a gopher had tunnelled next to the wall, and the tunnel acted like a funnel, causing the flood. Another letter told how the builder of a log-constructed dugout had used pitch to protect the earthside of the logs, then coated the whole structure with a layer of bentonite clay for waterproofing. He told how well the logs insulated the inside, keeping it dry and pleasant. Snakes were a problem to one builder. He lived in the western part of Nebraska where rattlesnakes are quite common. During extreme heat, the snakes moved into the dugout with the family. The builder's solution was to trade a pig for a neighbor's rat terrier that specialized in snake extermination!

Drifting snow blocked the south side of one dugout so completely that the family had to take turns digging a tunnel to the outside. The same letter included a poem written by the oldest girl of the family, who expressed the warm and secure feelings she experienced in the house during the storm.

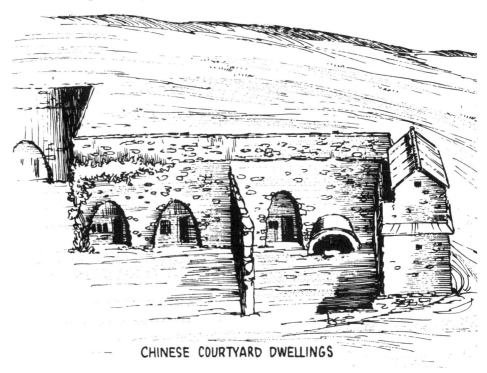

CHINESE COURTYARD DWELLINGS

On these two pages: views of various earth shelters built in Nebraska during the nineteenth century.

Photos courtesy of the Solomon D. Butcher Collection, Nebraska State Historical Society.

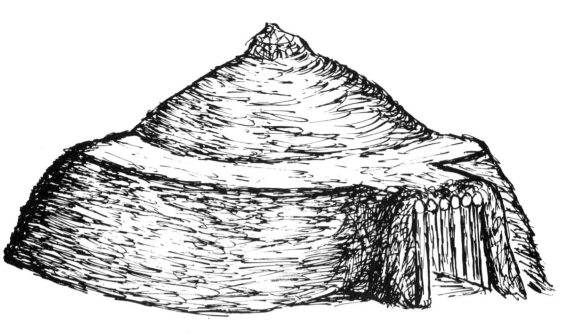

Indian earth lodge commonly found throughout the United States Midwest from around 1700 to 1885. The base of the lower wall was six to eight feet thick.

Interior framing of an earth lodge is shown above. All poles were tied with rawhide; the smoke hole was placed at the top of the structure.

A summary of this research led me to a number of conclusions. The inability of the various dugout builders to communicate with each other meant that each builder had to solve all of his own problems. This was a real disadvantage, but it did result in a wide variety of solutions to the various problems of building underground. Common problems seemed to be the same ones that underground builders face today. Waterproofing, drainage, light, support systems, insulation, adequate ventilation and the lack of certain useful building materials all concerned the pioneer builder.

Structural systems were usually post-and-beam, and utilized trees that grew along the rivers and creeks. Ash, hickory, walnut and hedge were the usual hardwoods found along the creeks and rivers. These woods were commonly used in post-and-beam support systems. The shallow "A" roof or the more common shed roof were the most widely used in dugout construction. These roofs were usually either shingled or thatched to provide watershed capability. As mentioned earlier, bentonite also was used to seal the surfaces.

It was near the turn of the century before enough small towns existed to supply a wider variety of building materials. The first extensive use of glass in dugouts to provide a more pleasant living space took place in the very early 1900's. Two things happened during that time: The larger assortment of materials helped the dugout builder approach the current state of the art (the social acceptability of these abodes also decreased as lumber became available for surface frame construction) and population was increasingly moving to the western United States. A summary of historical underground living leads to the conclusion that it is still one of the best ways of cooperating with the environment to create an economic, secure and energy-efficient shelter.

2
Preplanning Considerations

At present, the majority of earth-sheltered homes being built is located in rural areas.

Water/electricity

Most acreages, unless in a development, will have a well or cistern to supply drinking, bath, livestock and flushing water. One of the first things to happen in early winter in areas that freeze and have snow is an ice storm. These storms usually take out power lines for up to several days at a time, but you will not freeze in your underground home when this happens. Since a pump will not supply water without electricity, I usually recommend a gravity-flow water system, utilizing a windmill to pump water to a higher elevation and hold it in an underground or insulated tank. A small water tower that holds a thousand gallons usually works well, too, and has enough reserve for those days when the wind does not blow. I have had to pay to have water trucked in to my horses and animals because I did not think of this possibility.

Water lines should be placed below any possible frost line for the area you live in. Some areas near us have frozen more than five feet below the ground, and people have had to have water hauled to their homes until spring.

I believe in simplicity. That is why underground houses appeal to me. Many of the systems developed for country living have become complicated over the years in order to keep up with the city cousins. Any time an electric motor is installed to operate a critical item of survival, I start looking for an alternative. The windmill mentioned earlier is one alternative. If it is not possible to be completely self-sufficient in the area of electricity, a small wind generator and battery system to run critical items might be in order. The wind that is tearing down the power lines will give an oversupply of electricity from the generator, and a gas refrigerator will keep food during these times.

Site selection

Site selection needs to be considered on a broader basis than just the lay of a particular piece of land. Many people have a certain idea about the way their dream acreage will look. However, once you find that land, stop and look around before you buy it. Should an emergency arise, it is helpful if the house is located in close proximity to an area where help (medical, fire, etc.) is available.

The social environment will have to be considered along with the topography. Our land was exactly what we wanted, but we became very disenchanted with our life as it related to the surrounding community. The school district was not compatible with the educational needs of our children, and their progress was stifled as a result. Center-pivot irrigation was initiated on all of the land surrounding our property, lowering the water table and producing brackish water that was unfit to drink. The surrounding towns had no control over the development of the land in the area. The end result of this situation was the sudden explosion of tract housing near our property. The taxes went up. Everything that we had moved to the country for had disappeared within five years.

There are no guarantees against these things happening anywhere. There are ways of knowing ahead of time those things that are most likely to happen. Once you locate the land that seems most suitable, begin to analyze the total environment. I have broken the total environment into several categories which should be written on sheets of paper, with questions to consider listed under each of the categories. The main categories are legal description, political control, schools, history, neighbors, soil composition and structure, water, drainage, shopping, utilities, recreation, access and the labor–materials market.

Following are some questions that should be considered and information that I have found useful for each category; you may think of others.

Legal description

- Are there any liens against the property for materials, labor or services?
- Has the land been surveyed recently?
- Are the borders well defined?
- What are the mineral and water rights?
- Are there any grandfather clauses or other restrictions?
- If previous owners listed on the abstract can be contacted, they will be able to tell you about problems they encountered with the land or neighbors, etc.

Political control

This category has more ramifications for the owners of property than any of the other categories. Who will provide fire protection? Who is responsible for law enforcement and security, and how effective are these two services? Are there easements (utility companies, railroads, water districts, game commissions, neighbors, etc.) limiting your use of the land? We found out that the two railroads that bordered our land had control over any structure, fence or man-made object within two hundred feet of their track. That meant we had to get permission from both railroads to put up fences, utility poles, etc. Who are the political representatives (county commissioners, town boards, legislators) for the area? Are they controlled by special interests that might affect your location?

Taxes

- What is the tax structure, and how is it apportioned?
- What is the history of population growth of the area? Is it stable?
- Have tax increases been in line with population growth?
- Who are the taxes paid to (county, state, local government)?
- Is the land in a special assessment district?
- To which school district do the taxes go? If this district is an area removed from your land, what is the growth and need for new schools that will affect your taxes?
- Are there tax shelters (wildlife preserve, watershed, special use, etc.) that might benefit your land?

Zoning

- ◆ What is the zoning for your land? Is it agricultural, industrial, residential, commercial or special district?
- ◆ Is the land within the claimed jurisdiction of a nearby town?
- ◆ Is there a possibility of any town or development building homes, sewage treatment facility, landfill, storage, etc., on land that is adjacent to yours? Check your state industrial commission or planning board for possible industrial growth projected for your area.

Schools

- ◆ If you have children, what are their ages? Will they all attend the same school or will they have to go to different locations?
- ◆ Is school transportation provided by the school district?
- ◆ Does the type of system fit the needs of your children (special education programs, vocational learning, basic skills, student-to-teacher ratio, teaching materials, etc.)?
- ◆ Is the basic socio-economic background of the people and children in the school district compatible with your philosophy?
- ◆ What are the advanced educational opportunities in the area (vocational schools, state colleges, adult education)?

History

The history outlined on the abstract and the history that people in the area can relate can be of real value. Older neighbors can tell you about weather patterns, what has been planted on the land, the condition of the water, insects and pests that migrate into the area, and what noxious weeds seem to always rear their ugly heads. Generally speaking, find the large land owners and you will have located the people who most control the area. The neighbors who border your property will be of immediate interest. If these people are hard to get along with and exert a lot of clout, life can get difficult. The local tavern is a good information-gathering point. Don't ask a lot of questions at the tavern, just lead the conversation to the interest and scandals, etc. It will become evident what type of general attitude the area has towards strangers, small acreages, schools, politics and many other subjects.

Neighbors

We have discussed this under history, but a couple more pointers might be helpful. Get acquainted with the people who will be your immediate neighbors before you buy the land. Try to win their friendship and support. When an emergency occurs, these people will be your first help. Learn about their interests, pet peeves and whether or not they will at least be neutral, if not helpful.

Soil composition and structure

Let the land talk to you. A man I greatly respect, Emil Christensen, has let the land talk to him with more meaning over the years than anyone else I know. Emil Christensen is a retired architect who has spent his entire life developing small communities in rural areas. His belief that life is integral and should not be compartmentalized led him to develop communities for total living. His feeling that living, working and playing space should be congruent led him to adapt structures to the land instead of the other way around. His tunnelled camping shelters, brush shelters and carved fireplaces all let the land speak.

Your land is alive. Listen to it. It is filled with microorganisms, minerals and running water, and it will tell you all about itself. If there is no well on the property that is proven, have a well-driller from the local area give his opinion about the water table. Don't buy the land without a clause guaranteeing water. The well-driller will not only have a good idea about the water table, but he will also have knowledge of the soil structure of your property. This is very important since your earth-sheltered home will impose heavy loads beneath its footings.

The earth's shell is a series of overlapping plates that expand and contract as pressure at the core increases or decreases. Some of the elements that make up the earth's surface are as follows. *Faults* are breaks in the earth's surface that move up or down from internal pressure. *Seismic zones* are areas where earthquakes occur frequently. *Clastic mass* is a collection of pressed together rocks, and has joints and fissures throughout its mass. *Sedimentary zones* include most of the habitable areas. These are deposit areas resulting from glaciers, lake beds, marine deposits and the outwash of mountains. *Subsidence areas* are places that settle as a result of the loss of support due to the pumping of oil and water. California, Oklahoma, Florida and Texas all have this problem. *Volcanic areas*: Nothing needs to be said except that you *should not* build in an area with volcanic activity.

Most likely your land will be in a sedimentary zone. This means that the soil has been deposited in layers. These layers are not predictable and their composition will sometimes vary over an area of a few feet. The County Agent or Soil Conservation Service will probably have a detail soil map of your area. My own land varied greatly within two hundred feet of where I intended to build. The area that I intended to excavate for my house was mainly an expansive clay that was many feet deep, but the topsoil was only about twelve inches deep. Expansive soils in wet areas are more difficult to build on than in dry areas. The tendency for a building to sink is very great in this type of soil, and wide reinforced footings are vital.

Windblown areas many times will have soils that seem very solid and substantial. These soils are virtually cemented together by constant shifting and settling. There are records of people building on these soils only to watch their home sink out of sight after the cemented soil dissolved during a rain.

Heavy loads on any soil cause concern with its bearing capacity or ability to support the weight of a building or structure. The weaker the soil, the more it will be shoved aside as your structure heads for the center of the earth. This can be remedied by a footing system that equalizes and distributes the structure's weight over the excavated area instead of at its perimeter. Luckily, this problem is not as great for the underground structures as with surface structures. It is possible that the total weight of the structure underground is equal to or less than the weight of the soil removed during excavation. A "waffle" or grid type of footing system that is tied to the perimeter will give a very light footprint for the underground house, allowing construction on light or expansive soils.

Stability includes both lateral stability and expansive pressure. Lateral stability deals with cuts that have to be braced. Expansion is due to water absorption or freezing. Saturated clay has been known to exert pressures in excess of thirty thousand pounds per square foot. A nonyielding structure would fail.

Have soil tests performed over your land. This will tell you what trace minerals may have to be added in order to grow some of the plants that you desire. Look at the native vegetation that already exists. It is my feeling that if you aren't going to farm the land for produce, then nature will probably take better care of it than you will. A small amount of assistance usually will be needed in the reforestation or repopulation of plant growth on land that has been overgrazed, single crop-farmed or is badly eroded. Emil Christensen reforested thirty-nine acres of the most overgrazed, eroded and denuded land that I have ever seen, which became a virtual jungle in just five years. All he did was to encourage the growth of native plants that were already on the property by loosening the soil and adding humus. This land is now a park for the area, and the farmers whose worn-out land surrounds this oasis cannot understand how such lush growth occurs.

Percolation tests should also be run. These tests tell how absorptive the soil is. This will be necessary if a septic and lateral system is contemplated. A clay that becomes easily saturated will mean that extra laterals will be needed, or perhaps even an alternative system may be necessary. A preferable soil combination for growing or building is one that has a rock or mineral base at some depth, some clay above that to hold moisture for the plant roots, and a light surface soil with plenty of humus and nutrients for plant proliferation.

The ideal soil conditions for building underground would be diggable soils, flat slopes, bedrock below soils, seismically quiet, low water table and arable land for thermal efficiency.

Water and drainage

Ground water accounts for more additional cost to building underground than any other single factor. Near-surface water tables or underground streams might make it impossible to build. The cost for handling water of this nature would be prohibitive. Building near the top of a gentle slope by using a berm approach is the best way to get water to rapidly drain away from the structure. Since water always seeks the lowest level, springs and other ground water tend to seep out of the lower part of a slope, not near the top.

Shopping

This category is as vital as eating. The cost of fuel to run an automobile makes a shopping area located within a six-mile radius of your land very important. No matter how careful one plans or how large a garden one plants, the fact remains that additional food and grocery products are needed on a regular basis. It is surprising how many nongrocery items we buy at the supermarket. If there is no supermarket within a reasonable distance of your home, all of these store-bought necessities will have to be bought in scattered locations, increasing your gasoline expense. The cost of your land also has to include the cost of fuel, additional car repairs and inconvenience.

Utilities

We all should be moving towards energy independence, but the fact remains that you will probably have to be somewhat dependent on utilities, at least initially, unless you want to live with candlelight, no refrigeration and cook with wood or corncobs. If there is no electricity on the property that you are considering buying, it would be best to find out how much it will cost to have it brought into your home. Some utility companies are public-owned and are very cooperative. Others are either private companies or are state-owned. If you are planning to install hydro or wind generators as your primary source, check with the power company regarding their regulations on this equipment if it will feed into their lines.

Recreation

Fuel costs have forced people to look closer to home for recreational activities. Recreation and cultural outlets will not cease to be important to you in the country if they are presently important. Winters may even increase the importance of these outlets in easing the cabin fever that may set in. Check into surrounding towns for theatres, skating rinks, hobby outlets, parks, bowling alleys, etc.

Access

This term applies to more than a guaranteed route onto your land. Sometimes people purchase a piece of land that is surrounded by property that is owned by another party or division of government. Sometimes right of access can be denied by relatives of a deceased landowner. The owner may have been friendly and made a verbal assent to guarantee access. The relatives may not feel the same way and may try to squeeze you out in order to buy your land to make theirs more salable. Even years of use of a given driveway or road that crosses land owned by someone else does not necessarily constitute an easement. Check with an attorney regarding right of access before you buy land.

Access also denotes condition and maintenance of the roads you will be using. Rural roads can be difficult to travel at times: If they are dirt or gravel, rain can turn them into swamps that are impassable. In colder climates, snow may drift and block even the best-paved road. The county maintenance crews many times will wait until it is no longer snowing before plowing roads. This may mean that you will stay home and miss work, or you may have to stay where you work until the roads can be made passable. Families can become separated for days in some cases.

Labor–materials market

This part of the preplanning deals directly with the cost of the underground house that you hope to build. It is best to inventory the kinds of materials that are available within a reasonable distance before designing your house. The design may be dictated more by available, cost-effective materials and the skilled labor market than by your design ideas. I usually advise my clients to check around for builders and skilled craftsmen before doing any shopping for materials. If you are going to act as your own general contractor and subcontract the work on your home, the craftsmen that you hire will either make it a pleasant, fulfilling experience or one that you will regret. It really is not as important that they have experience working in underground houses as it is that they are conscientious and compatible with your personality.

Areas that are somewhat remote may also be short of skilled labor. Work crews that have to travel any distance usually start charging from the time they leave their shop until they arrive back again at night, which can considerably add to the cost of a structure. Check with farmers in the area who have had a building constructed within the last year. They usually know who the good carpenters and masons are. Talk to these craftsmen on a personal basis and get their gut reaction about underground building. If they seem enthused, this is a real plus. Acting as your own contractor usually means that you will need to have a full-time foreman on the job that you can trust. If you can take three months off from work or if you are self-employed, you can act as your own foreman. Constant supervision is necessary to assure that everything is coordinated and done the way you want it done.

Acting as your own general contractor can save you up to twenty-five percent of the building cost. You had better be prepared for the headaches as well as the savings. There are three enemies in the building business: weather, scheduling and money. One of the first things to happen is a very heavy rain just when you are ready to pour footings. The footings not only fill with water, but many times the excavation will have to be braced to stop the walls from falling or sliding into the work area. This can go on for a week or more, throwing the scheduling into complete chaos. Many jobs are interconnected, and one cannot be performed before the other. This means rescheduling a whole series of jobs and people. One or more of those people who are subcontractors may have their work scheduled for several weeks, and when you take them out of your schedule they may have to later move you back several weeks in theirs. This requires an unreasonable time frame for building, and you will have to find a substitute for that particular subcontractor. Buying your land in an area that has a good skilled labor market becomes very important at this point.

The skilled craftsmen in the area are usually indicative of the types of building materials prevalent in that area. An abundance of masons usually means either a good block plant or brick factory in the area. Carpenters are usually plentiful, since lumber is available either at local sawmills or through lumberyards that buy in large enough quantity to keep this commodity competitive. Plants that provide ready-mixed cement are also found everywhere, but the craftsmen skilled at poured-in-place construction may be hard to find. Cement finishers are not necessarily good at setting up forms that require precise placement of reinforcing steel. It also may be better to pump these forms to assure even distribution and consistency, a process which requires specialized equipment.

Check to see what prefabricated materials are available. Sometimes, if there is a large enough population in the area, a cement plant that sells ready-mixed cement may also pour precast roof panels, twin-tees and tilt-up wall panels.

Steel-building franchises offer still another viable building product. Steel trusses can be used to support steel sheeting that will carry substantial loads. Steel arches and culverts can also be incorporated into a design.

Pressure-treated lumber also offers a viable option. In some areas, a post-and-beam system covered by pressure-treated sheeting is preferred over other more costly materials. The subcontractors will advise you of good materials, but do not buy on their recommendations alone. They will want to sell you a material with which they have familiarity, which may not necessarily be the best material for your construction design. Do comparative shopping for labor and materials. Asking questions is your greatest protection. Many craftsmen will overstate their abilities. Ask them for a list of their clients, and call a few of them to see if they were satisfied with the work and the final bill. When shopping for materials, ask about the shortcomings of a product, as well as its positive qualities.

Manufacturers usually state only the positive things their products will do. For instance, if you are buying a waterproofing material, the instructions will tell how to apply the material and how good it is at keeping moisture away from the building. What they have not told you is how alkali that may be present in the soil that surrounds the building can leach the waterproofing material. The soil is a veritable mineral repository. Many of these minerals will interact with the building products with which they come in contact. If a given product does not state how it will react to the minerals that are found in your soil, it may be necessary to call the factory and visit with their chemists to get their reactions. If the company does not know how their product will perform under the conditions that you want to use it, do not buy it!

Some seemingly harmless products can be deadly under certain circumstances. For instance, certain plastics under heat release deadly gas and PCB's that can cause everything from memory loss to death. Ask questions about the safety of the products under the conditions that you intend to use them. Plastics, insulations, preservatives, waterproofing compounds, sealants, any combustible materials and other products that will come in human or animal contact should be carefully evaluated. Some urethanes continually give off gases that are harmful; these should never be used in areas where their fumes will be inhaled. Some of these "miracle" products have their places and are very effective, but they should be used only under the conditions for which they are designed.

Indigenous materials can save you a small fortune. Do not get locked into some design that limits the use of these materials. A cord permit can be obtained in some areas at little cost and allows you to cut trees for all of the posts, rafters, beams and braces. These in turn can be cut to size at a local sawmill for a fraction of the cost of processed lumber at the local lumberyard. This also allows you to be more selective in the grades of wood that you cut. Cordwood, cut and dried for burning fireplaces, also makes a great-looking—and thermally efficient—wall. Native stone can be used in place of brick and other manufactured material. Even direct earth-tunnelling can be an attractive option in some circumstances and will completely eliminate the need for a structural shell (see Chapter 19). Where expansive pressures do not have to be considered, rammed earth is another consideration. Adobe has been used very successfully in several underground houses in New Mexico.

Recycled materials should be at the top of the self-builder's list when shopping time arrives. Two days after I bought "I" beams for my first house, I found out the exact beams could have been bought as scrap for much less from an old store that was being demolished just a few miles away. The tendency is to buy new material and get the buying process over with so the building can be completed as quickly as possible. This is one reason I tell all of my clients to start shopping for recycled material a year or two before they actually intend to build.

Steel is not the only possible recycled material that you will want to consider. Two by fours and other dimensional lumber that will not be exposed to view can many times be bought through salvage yards and demolition companies. Not too long ago it was not economically feasible for a wrecking company to pay for the labor in order to salvage wood, glass or other material. The cost of new material has now created a new breed of salvager. The salvager buys the rights to the material he wants in a given building from the demolition company that holds the demolition contract. This subcontractor is given a certain amount of days to remove these materials before the wrecking ball does its job. These entrepreneurs usually advertise their wares through local papers and specialty classified papers. Stained glass, other decorator items and unusual hardware can be purchased at a fraction of the cost of new material. Light fixtures and doors add many hundreds of dollars to the cost of a house if bought new. Glass is also an option for some greenhouse applications. Wiring and plumbing should be purchased new, however, since these items tend to self-destruct with age.

Many people have asked me about ordering new materials and stockpiling them to hedge against inflation. This question has no simple answer. Each area has competitive situations that vary between suppliers. Generally, I find it best to prebuy strategic materials such as steel and some prefabricated concrete products. The difference between what it costs for interest on the borrowed money used to buy these items and the inflation increase in price when they are bought at time of construction usually favors prebuying. As stated earlier, prebuying recycled materials is always a good idea.

Other considerations

Ask yourself about how long you intend to live in that particular home, and how many people there will be in your family during that time. It is a fact that most people live in a home for an average of five years. This is rapidly changing, and the country is becoming less mobile. If you plan to live in the location for only a few years, ask yourself who will buy the house when you move. Financial institutions are only interested in resale when considering a loan on a given structure. The resale value of any home will be determined by comparing what houses of similar nature are bringing on the current market. Since there are not many underground homes that have been sold, there are few comparisons a financial institution can make. An analysis of what size and cost range will be in demand by the buying public should be made before you design your home. This will make it easier for you to convince a mortgage company to lend you the money you need. The trend in home building will be towards less square feet and more compact energy-efficient design.

A good way to avoid problems with family members when designing your home is to include all of them in the project. There is often a tendency for everyone to want more space than can be accommodated for within the size of the structure you may be able to afford. Floor plans are very difficult for most people to visualize in relation to the actual use of space.

Difficulty with spatial interpretation can be minimized in two ways. One way is to draw each room's actual size on the ground. Place some furniture in this space and have everyone move around in the space. Walls give an added perspective but are not necessary to get the feel of a space. The second way to establish functional space is to build a half-inch scale model of the house. Architectural supply stores or hobby shops will have scale furniture, plants and people to aid in arranging the floor space, doors, storage, etc. This process will aid in the compromise that will be necessary to keep everyone happy with the end product.

One additional precautionary note: Don't become discouraged when the house is "staked" out by the surveyor. The house will continue to look too small until it is closed in with the walls and roof. This size and space perception is very misleading.

3
Permits, Insurance and Financing

Permits, rules and regulations are basically designed to protect the buying public. Having been on both sides of the fence, I recognize the problems as well as the benefits of these regulations and the people that enforce them. The good side of the regulatory maze is that its main purpose is to eliminate "buyer beware," insure basic structural integrity, adequate ventilation, proper lighting, reasonable sanitation, minimize fire hazards and insure general safety.

Permits & zoning

Your first stop before putting a down payment on the land of your dreams should be the building permits and inspection division of whatever branch of government has jurisdiction in your area. The rules and regulatory bodies that you may have to deal with are broken into three areas: the Unified National Building Codes (used as a standard by most legal authorities), building and housing codes adopted by various cities and communities to meet local or regional regulations not covered in the Unified National Codes, and neighborhood covenants that are additional rules applied only to a given neighborhood.

Generally speaking, in most areas, if you own twenty or more acres, you can build pretty much as you please. This requirement has been raised to forty acres in some areas. You most probably will fall under the jurisdiction of the county in which your land is located. The county may have its own permits and inspection division, or it may utilize the facilities and people of the largest city within its borders.

Even if a building permit is not required, you probably will have an electrical inspection and must meet basic fire laws of the county. The state fire marshall has inspection privileges in some instances. If you are located in a rural area, the county health department will also have jurisdiction over what type of waste-handling system you can install. Depending on your location, you may be dealing with state, county, city and local officials at any given time during the building process. Unless building underground is specifically prohibited in your community or county, you will be dealing more with compromises than trying to change rules or getting exceptions.

There are some communities that have specific regulations limiting the type of structure that may be built in a given location; this falls under zoning. Most cities, incorporated towns and some counties have zoning ordinances. By and large, these ordinances are good since they insure that disparateness of building types and uses is kept to a minimum for a given area (this keeps heavy industry from building next to a housing development, etc.).

After World War Two, a number of communities adopted special language into their building codes that now has the net effect of limiting your ability to build an underground structure. This language was initiated to deal with the phenomenon of "basement" houses that were built by returning veterans. GI loans were obtained by many of these men for the purpose of building a home. Basements were completed initially to provide emergency shelter for their families until the main structure could be built. A large number of these families continued to live in these basement houses without completing the above-grade structure. The end result of this was unsafe, poorly lit and poorly ventilated living conditions that were not only a hazard to the occupants, but were also a visual blight to the area in which they were located. Depending on the language used, below-grade structures were banned to obviate this unpleasant structure from the communities involved. If this is the case in the community or jurisdiction where you wish to build, you may have to either take legal action to remove or change the language involved, or get an exception.

The area where I built my first underground home fell under a language stipulation much as described above. My design and blueprints were all in order and met the general building codes for the area, but legally I could not build. This distressing situation was discovered long after I had purchased my land and lived on it in an old existing farmhouse. At this point, let me say that the best thing to do is not become belligerent with the authorities because you may then have difficulty securing a building permit and may possibly have your land condemned. If your land is condemned, you will probably not only have difficulty getting the condemnation removed, but it would also be difficult to sell the land.

Depending on who has jurisdiction, there are several approaches to the problem. The first approach is to find out what would be required to get the zoning ordinances changed to allow you to build. Let me state at this point that it is the job of the permits people to help you get a permit, not the other way around. If you are cordial and cooperative, they probably will be, too. If a change of language is needed in the code to allow you to build, it could take up to two years for the process of events to unfold. In my case, we found out that it was the City Council that would have to make the decision.

We hired an attorney to word the addendum, then hired a lobbyist to get it on the Council agenda. It took eight months to get it on the agenda. The Council assigned a committee to study the wording and make any suggestions they felt appropriate. After they were satisfied, it went to the City Attorney to assure that it did not conflict with any existing ordinances or laws. At the end of the second year, the Council adopted the addendum to the Codes that allowed the building of underground structures. By the time the ruling had passed, I had been living in my underground house one year. The reason I was able to build without waiting for due process was that I was able to obtain an exemption.

Some additional tips on how to gain popular support for the long-term changes in building codes and covenants might be helpful at this point. Home builders associations and chambers of commerce can become your allies, but they may take some convincing. Work up a good slide program and present it to these groups and other civic organizations. This educational program will benefit the entire community. Get your facts together regarding how thermally efficient these structures are; the slides will show the attractiveness of these structures. Use some of the large commercial buildings that have been built underground for precedent. Mutual of Omaha has a 190,000-square-foot underground office in Omaha, Nebraska, which is topped by a ninety-foot diameter and a thirty-ton dome. The beautiful underground Terraset Elementary School in Reston,

Virginia, has very impressive energy figures. The school was designed by Douglas Carter of Davis, Smith, Carter & Rider, Inc, and a 16 mm motion picture is available through that firm regarding the school's construction.

Special districts

Another way around some codes is to establish a special district under the zoning ordinances of a city or area. This special district would include all underground buildings, either homes or commercial structures. Many cities are going to special districts exclusively in order to establish a more cohesive building plan for their city. Special districts have been established to protect the architectural integrity of historical areas, etc. This would really benefit underground building since it would guarantee solar access for heating. A tall building erected next to an underground structure could shade it completely.

When a town planning board or other officials learn that an underground district could be a veritable park that would actually contribute more useable space instead of remove it, they will become more interested. Also, all of the grass, shrubs and trees will help clean the air and add oxygen, as well as serve as accoustical barriers against traffic noise. These facts, coupled with the ability to drastically reduce consumption of fossil fuel, should be powerful persuaders.

Use of the special district also alleviates the problem of having to adapt the overall zoning code for coverage of subsurface development. San Francisco, California, has gone one step further and instituted a new land-use control system that places all land in special districts.

One of the things that may also be in favor of creating special districts for earth-sheltered construction is the environmental-impact statements being required by the Environmental Protection Agency before any major construction projects can be implemented. Earth-sheltered structures will not conflict with the visual appearance of the area and will add green space to the inner city, creating a parklike atmosphere. The added incentive of less physical and psychological damage by noise and visual pollution should also be considered.

Inspections

Let's look at the individual sets of rules that will govern permits and inspections. The Unified National Building Codes have been adopted on a national level by most states, counties and municipalities as a basic guideline. These amount to several hundred pages of specified rules. Cities and counties add their own particular rules to this body, then neighborhoods add their covenants to which you have to agree if you want to build and live in that particular neighborhood. Meeting the Unified National Codes and the city ordinances are not all that difficult, other than the cases previously mentioned. There are some requirements that may be more difficult to meet than others, however. In some cases, it may be difficult to meet the egress provisions that state all habitable rooms must have two exits, one of which must be directly to the outdoors. Without making the design look like a motel with all bedrooms located in the front portion of the structure, several exits through the back utilizing tunnels might be necessary. This could be costly as well as impractical and would not necessarily be the best answer to window exit provisions.

Having visited with several fire department chiefs, I am convinced that they would favor fire-protected corridors and smoke alarms as opposed to window egress only. Invalids, the elderly, babies and those who are incapacitated will find a window of little use in escaping a fire. What is needed is an alternative route out of a structure that is fire-protected and has a positive draft system that draws smoke and fire away. This would allow firemen to evacuate everyone safely from the structure. Statistically, more people die from asphyxiation and smoke inhalation than from flame. This argues most favorably for the installation of smoke alarms. These two systems are superior to window egress. Once the local fire chief's support is gained, it usually is not very hard to get an exception granted.

Red-lining

Red-lining, though illegal, can still be found within some communities. This practice consists of designating certain areas as less desirable than others, usually based on income level or ethnic population. The reluctance of mortgage companies to loan money for housing in these areas is the best indicator of this practice. If this seems to be the case in the area where you wish to build, I would file a complaint with the Federal Housing Administration and the Department of Health and Human Services.

Neighborhood covenants

Neighborhood covenants are the most discriminatory and difficult to deal with when trying to get a building permit within a metropolitan area. Some particular covenants deal with how many square feet you must have in the home, basic architectural style and even how your lawn must be maintained.

Performance codes vs. restrictive codes

A last thought regarding creating long-term change and impact on the building code structure: After reviewing most of the existing codes, I have come to the conclusion that any additions or corrections made to them should have a different base. The majority of building codes are either restrictive or stipulated. This greatly inhibits creative approaches to solving certain building problems.

It is my feeling that any changes in the codes should be performance-based. The difference between a performance-based code and one that is restrictive or stipulated is that the latter requires numbers, types of material, and limits creative change that might produce a more sound and economic home. Performance-based codes set a performance standard for the products, techniques and methods to be utilized, but they do not designate those products, techniques and methods. This would encourage innovation while still protecting the buying public. There are new materials and techniques being developed every day, some of which are far superior to what is stipulated in many cases but they cannot be used. The fire corridor idea is one innovation. Skylights, light wells and other methods of bringing natural light to the interior of a structure in lieu of windows are other innovations. A composting toilet in place of a sewer and septic system is another idea that would take some of the strain off the environment and the taxpayer's pocketbook.

Insurance

Insurance might not seem very important until it is needed. Depending on where you will be building, your insurance liability will vary considerably. If you are a professional builder, you will have to look at underground construction in a little different light than surface construction, mainly because of its unique nature. Until underground buildings become more common, you will be confronted with large numbers of curious people. These people, by and large, will be very friendly, but the fact that the building may receive more publicity than normal will draw the curious in larger numbers. This creates a hazard for both the self-builder and the building professional. I had over two thousand people on my construction site and about five thousand over the next two years after construction was completed. I took additional liability insurance to protect myself and the interested visitors.

You become liable for people who may climb into, onto or over the structure and excavation (as a swimming pool owner is liable for anyone who drowns in the pool, even though it may be protected by a fence). My advice would be to check your liability and then take extra precaution to block access to dangerous areas where material is being moved or digging is taking place, etc.

One additional thought about insurance. When excavating in the city or heavily populated areas, the excavator and the builder have to assume liability for settling, cracking of foundations or other structural damage that occurs to buildings adjacent to the excavation.

Financing

Financing becomes very essential at this point. After applying for a mortgage at several of the usual retail outlets such as banks, mortgage companies, etc., you may discover that they are not all willing to loan money for underground building. This will be partly because they are unfamiliar with the subject, partly because they have very few underground homes that have been sold to use as a yardstick for measuring resale, and partly because the mortgage market is depressed due to high interest and tight money conditions.

Rural banks sometimes are a good source if they have money available for mortgages at all. If you live in the area, local banks are required to lend a certain amount of their capital for projects in their customer area. Make a selling presentation using your house model, drawings and the preplanning material just covered. This will let the banker know that you are knowledgeable and have done your homework. Sometimes it is best to go for a construction loan, then try to give the banker the opportunity to see the finished product, thus reducing his skepticism.

Mortgages are divided into about five types: commercial, industrial, residential, farm and ranch and special-purpose properties. Under special-purpose properties, you have to have a special purpose or use to which you would put the building, and you pay a higher rate of interest under a shorter term.

Periodically, funds are not available on a large scale from traditional sources such as savings and loan companies, life insurance companies, banks and trusts, commercial banks, mutual banks and mortgage companies. The Federal National Mortgage Association, the Federal Home Loan Board and the Government National Mortgage Association are all federally funded programs that receive through Congress funds which are available to builders/buyers of above-ground structures. A lot of their money is available under specific programs that have been established. However, there is no program presently in existence that supplies money through these programs to the retail mortgage market for underground construction. It will be up to the self-builders and professional builders, along with building and trade associations, to lobby for such funds. Write to your senators and congressmen and encourage the various associations to lobby.

Possible sources for financing you should explore are: credit unions, life insurance policies which have cash value that can be borrowed against, and second mortgages on land; you can buy land, subdivide it, and then sell building sites specifically for underground homes, using this profit to pay for your home. Another alternative is to form a corporation and sell stock in a development of earth-sheltered housing.

4
Purchasing Land

Finding the land that you want will probably be more difficult than designing the house you like and can afford. Your family and personal needs often conflict, making what you as an individual would consider the perfect site, an unusable one. Being a single parent with three daughters makes my choices even more difficult. Diversity of age adds further complications, since my two elder daughters are pursuing their educational goals at two different schools and vocational levels. The youngest attends grade school, which means that I have to be located, at least for the next few years, near suitable school systems or assume an additional expense of board and room.

Chapter 2 already discussed the necessity of examining the possible building site from the total approach, so I will talk mainly about how to find the land that most nearly meets all of your requirements. Now, since I cannot possibly know your exact needs, I will cover some hypothetical situations. Whether we like it or not, your land requirements and building specifications will most likely be primarily determined by economics.

The ideal situation for a young couple or individual unable to secure a loan from conventional sources is to possess a hobby, skill or ability that would provide the residual income necessary over and above the basic homesteading returns of food and shelter. This would allow land-shopping in remote areas or areas of less desirability for those who have to work a regular job. This fact automatically reduces land costs to nearly realistic values. Writers, artists, potters, silversmiths, singers, musicians, handymen, mechanics, nurses, therapists, teachers, carpenters, plumbers and masons all have skills that usually are in demand almost anywhere.

Rock quarries

If your situation demands that you shop for land within a reasonable commute to work near a populated area, then there are several places to consider. Abandoned rock quarries can be found in and around most towns and cities of any size. One of the beneficial characteristics of earth-sheltered construction is that it creatively utilizes land that is otherwise worthless. Some of the old rock quarries have been allowed to revegetate, and some even possess lakes at their lowest levels. The rugged beauty of these old excavations lends itself to imaginative design and planning. The only drawback to some of these quarries is that drilling for water may be difficult or downright impossible. The advent of county water districts may provide the alternative solution needed to make the quarries viable building sites. The rock base underlying most of these diggings nearly eliminates the need for footings. Excavation, other than for backfill, may not even be necessary.

Abandoned rock quarries can serve as economic and attractive building sites.

There may even be some federal funds available to help with the reclamation of this land through the Environmental Protection Agency. Contact your congressional representative's office to check on funds that may be available. Application for these funds will be enhanced by a comprehensively written, and if possible, illustrated plan. Developing one of the abandoned sites with five or ten low-income underground homes might be a good way to have your own paid for through a project management fee.

Railroads

Another avenue to explore when trying to locate a building site near a center of population is to check the railroads in the area. Once in a while an abandoned right of way is put up for sale. Some of this land goes at public auction, while some of it is sold by bid through the railroads. I have seen several desirable locations sell for very little money due to their poor location for tract-house development and industrial purposes.

Many times farms have corners or heavily eroded areas that are not suitable for anything other than growing weeds. Some of these could be ideal sites for underground dwellings. One of the problems here is that it becomes very hard to legally subdivide into less than twenty-acre plots. A way around this might be to secure a long lease with irrevocable terms. There are many areas that might be just what you need near your present location; you just have to let people know you are looking. Real estate companies and banks are not your best bet, since it is more profitable for them to engage in large transactions. The Soil Conservation Service, Co-ops, granges, county agents, well-drillers and repairmen, county linemen, county deputy sheriffs, Izaak Walton Leagues, state fish and game officials and fishing or hunting friends will all be good resources to aid you in your land search.

Government land

Land sometimes can be purchased at a tax sale. This does not happen very often, but once in a while a piece of good land can be bought for the amount of the taxes that are owed on it. Check the county tax records, and use every avenue that is available. Government lands become available from time to time and may be sold at reasonable prices if the desirability of the land is low for development. Abandoned missile sites have been bought at very reasonable prices in Nebraska. Several of these have been converted to underground living space. A large portion of a missile site is already underground and may only need some imaginative adaptation to become habitable. Federal funds may be available to help you convert this space under reclamation funding. The old Nike antiaircraft missile bases are good prospects since they did not have deep vertical silos like the inter-continental missile sites. Ammunition storage bunkers make great conversions.*

Sources of free labor

Vocational and technical schools are good places to find willing hands and creative minds to help you build your underground home, as is a community college system. Many times apprentice carpenters, masons and other skilled help can be recruited for class credit from these institutions at a fraction of regular scale.

Buying farmland

I will close with a few additional thoughts regarding purchasing and financing land. If a farmer is willing to sell a piece of land and you can get around the subdividing problem, a land contract may be a good deal for both you and the farmer. The land contract allows the farmer to avoid an increase in taxes by spreading the capital gains over a number of years, while still having an increased regular income from land that would otherwise be nonproductive. Several of my friends have bartered their skills and labor for a couple acres of land. This is made possible sometimes when several farms or pieces of land have been combined into one farm. Sometimes the farmstead or house site on one of these pieces of land can be deeded separately.

Creating a park

One last thought about obtaining land: One of my friends bought some very badly eroded and barren land with funds obtained from a community fund for creating a park near a small town. He got the funds, even though the park would be a privately owned one. He had a written agreement that he would open the land for parties and camping upon prior notice to supervised groups. The result was a really beautiful twenty-acre park which was developed over the years. The primitive underground home became an integral part of the park's pioneer appearance. The house was built from trees on the property and was a replica of an early dugout in the area, and the townspeople even established a fund to help with maintenance. The relationship has developed into a profitable one for all concerned. People outside the community area pay to visit the park, and the funds generated are shared by the town and the park. Many small towns with a sense of history have established historical societies or committees that one

---

*A number of bunkers have been turned into factories and small businesses near Hastings, Nebraska.

could approach with a similar proposal. The possibility of obtaining state or federal money for a project of this nature might be very good, though I have not investigated this possibility.

A large number of farms are owned by absentee landlords or corporations. A profit-making park with you as manager might appeal to these absentee owners, since almost all large farms contain land that is marginally profitable at best. By using twenty or thirty acres for a park, the corporation could take a tax credit for aiding in the development of the ground for this purpose. Once the park became profitable, they could then either use the profits or make it a self-perpetuating nonprofit corporation. In any case, it would be a good tax write-off and would earn the donor valuable favorable publicity.

Depending on the terrain in the park, tunnelled camping shelters with fireplaces and sleeping bunks could be created as an added demonstration of underground utility. A popular display on the *Mother Earth News* compound has been the tunnelled camping shelter that I designed and helped dig. The comfort, security and seclusion is fascinating, and when visitors realize that they can do the same thing at little or no cost, they can hardly wait to get home, find a wheelbarrow, mattock, spade and shovel and get started! This design will be covered later.

Funding alternatives

Funding any purchase of land is a monumental task if sufficient cash or real estate is not available for use as collateral. Co-ops formed with friends can often pool enough money and resources to make the down payment on a piece of land. Some of the land is just right for five or ten economic underground homes. The land can be owned as a single tract and leased in a way that is similar to leasing condominiums. A corporation can be formed instead of a co-op. This gives the potential for selling stock and returning the stockholders' money with good dividends after selling the homes and the developed land. Hopefully, this will accrue enough money for you to own your home and land out of the development. Many times innovative ideas such as these are not tried because of the time it takes to promote and sell the stock or sign up co-op members.

Any of these ideas should be thoroughly researched, and a very graphic presentation should be made. A relief model of the topography, complete with a model of the proposed dwelling and miniatures of plants, would be most graphic. As I previously stated, an architectural student or drafting student might like to participate in the project for class credits. State historical societies, state parks departments, city park and recreation departments and private philanthropic organizations are all to be included in initial research for financing.

5
Basic Earth-Shelter Modes

Designing and building an underground house is a little like designing an airplane: After all the calculations are on paper and the design is put into test form through models in wind tunnels, etc., the time comes to build the actual aircraft. Many design changes occur during the actual fabrication. When it has been ground-tested, the real moment of truth comes when a pilot tests it in the air. Having designed airplanes as well as underground buildings, I can tell you the steps involved are a lot the same. After you have designed, built and lived or worked in an underground structure, you realize that most of your preconceived ideas were bent in the direction of your prior experience with surface structures.

Having taught the underground building seminar for building professionals at *Mother Earth News*, several things have become evident. The tremendous amount of interest on the part of the building professionals made me aware that underground building was not going to be just a fad. Having been consulted on the design and construction of several hundred underground buildings, mostly with self-builders, I have sought to compile information that will reduce most of the problems that arise into more manageable proportions.

One of the first questions that I am usually asked is, "Do I have to have a south-facing hill or slope on which to build?" Rather than providing a definite answer, I would like to illustrate the three basic modes in which an earth-sheltered structure can be built and then explain the pros and cons of each, letting the self-builder decide.

There are three basic modes for constructing an underground building: bermed, envelope and a slope cut, and they depict the type of terrain that can be utilized for construction. As the illustrations show, these three categories represent almost any imaginable topography. Actually, I try to encourage people to shop for terrain that cannot be used for productive farming, since millions of acres of productive land are torn up for tract house development every year. Some areas have no hills that would allow a south-slope cut, so the individual is confronted with having to use flat ground. This type of topography can be built on just as easily as the slope. Either the bermed or envelope structure can be used. Some locations that offer only flat land also have high water tables; it is wise to consult a local well driller in any case. If you have a high water table, you will be forced to use the bermed approach. Houston, Texas, is one city where bermed structures are the only way one could build an underground structure.

Bermed mode: advantages

The bermed mode is the most adaptable mode. The rectangle seems to adapt to the largest number of sites when used with a berm. The bermed mode can be used on flat ground with high water tables, on a slope with only a shallow cut

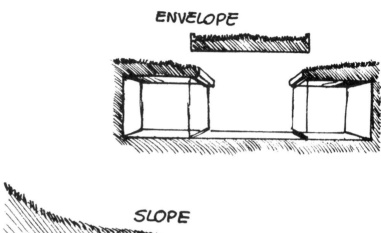

Shown above are the basic earth shelter modes that all underground structures utilize, depending on the terrain.

necessary, and in populated areas, creating a parklike atmosphere for the home. The bermed mode allows the use of the land with the least amount of excavation. When this mode is used on a slope or hill situation, the structure should be located near or at the crest. Earth is then brought up and around the structure to earth-shelter it. This has several distinct advantages. The top of the hill offers the best view; by not making a deep cut or excavation, there is less need to deal with expansive pressure. In a cold climate, the berm allows the soil to heave without all of the force being directed into the walls of the structure. Since there isn't the mass of earth behind and around the bermed structure, as in the other modes, the earth can expand and contract without this mass directing the force in only one direction.

When building in a populated area, such as a city lot, the bermed mode allows the building to be above ground, yet still provides the privacy for which underground building is noted. Proper design of the berm in association with terraces and visual planes will create a garden atmosphere that will accentuate nature and hide the house. Light and ventilation are also easier to distribute in the bermed rectangle. The ability to build underground and above the ground in the city means that there will not be difficulty in using sewers and other utilities in a conventional mode. Since earth-sheltering may be looked upon less than favorably by the neighbors, the above-ground bermed mode may also fit in better with surrounding architecture. Creating the illusion of a garden that makes the house disappear will be illustrated more fully in Chapter 21.

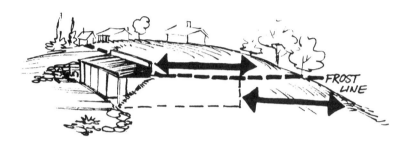

The berm of the Prairie Cocoon transmits frost heave and expansion in two directions, thus reducing structural stress.

Envelope mode: advantages

Presuming the water table is not high and you happen to like the envelope approach, there are a number of good points in favor of this mode. Esthetically, of the three modes, the envelope changes the natural landscape the least. If the building will be located in a very cold climate, this mode is completely isolated from wind chill, which in colder climates does more damage than the temperature. A number of commercial buildings have used this mode very successfully, as have various homes. Rectangular, square or round structures adapt to the envelope mode well. Paul Isaacson's round house (with its double dome covering the atrium) is a good example of the envelope. Mutual of Omaha's three-story square building, which uses a ninety-foot diameter dome to cover the central garden cafeteria, illustrates the successful commercial use of the envelope.

The envelope allows the ground over the structure to be used for a variety of purposes. Some of my clients garden this area. Paul Isaacson used the space between the outer dome that surrounded the entire roof area and the inner dome which covers the atrium for a garden. The structure's vulnerability to tornadic winds or even a nuclear blast is almost zero. The inner court or atrium can also become a lush tropical garden that can be enjoyed year-round, regardless of the climate. Natural light can penetrate the rooms surrounding the central open area. This is a distinct advantage in the summer when the sun is more vertical in most areas. The envelope is also entirely isolated from noise pollution, which makes it an exceptionally good mode for inner-city construction. The view of the courtyard is far more rewarding than heavy traffic or the neighbors' burning hamburgers on their barbecue. Envelope homes constructed around large airports would protect their occupants from the nerve-rending noise levels.

Slope mode: advantages

The slope cut is the mode with which most people seem familiar and was also the most commonly used by the pioneer dugout builders. The reason for this was that it provided for the easiest excavation since the dirt can be brought directly out of the face of the slope and does not have to be "tracked" up an incline or brought from another area for backfilling. The slope cut also makes for easy drainage of water away from the structure. I always recommend the French drain field coupled with footing drain tile, no matter how positive the drainage is. The slope cut allows the drain tile to exit at either end of the front of the structure. This eliminates the additional absorption field or a connection with a storm sewer.

The view from the slope side of the structure is excellent and can be increased in radius through the addition of an extruded greenhouse. This greenhouse is not only an attractive addition to the structure, but it will also serve as a plenum for air exchange and growing winter produce. Natural light is readily available and will provide solar heat without interference in the winter.

Just about any shape of structure will work in this mode, although the rectangle is probably the most easily adapted. A wider structure that allows more glass exposure will create more solar gain for heating the structure in cold climates. A structure that is wide with limited depth will admit natural light to the entire building without resorting to an excessive number of roof openings for skylights or light shafts. These additional openings are not only costly, but they also create another possible leakage problem. Freezing and thawing take their toll by cracking retainment surfaces with which heaving soil comes in contact.

The slope cut offers a lower profile to north and northwest winds in cold climates. Once again, the slope cut tends to blend more with the surrounding terrain, maintaining more visual harmony with the environment. Natural air movement through windows and doors is possible with the slope cut, although I would still install a soil pipe and solar chimney system (see Chapter 7).

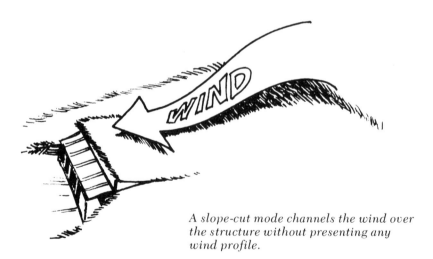

A slope-cut mode channels the wind over the structure without presenting any wind profile.

As a partial answer to whether or not one needs a south-facing hill to use the slope cut mode, I would say that it is not necessarily always the case. The south-facing slope has the advantages discussed, but there is a way to use a north slope or other orientations. The method involves the creation of a semi-envelope. The cut is made in normal fashion, but the dirt is brought forward to create a dike or berm. This berm then becomes the back wall of the structure, and the front of the structure will face uphill, using the deepest part of the cut or excavation as a courtyard or atrium. The excavated soil will be used for a berm around the three sides of the structure, with the glass wall of the structure opening onto the courtyard. As you are probably beginning to realize, there are very few absolutes, but rather various means of meeting existing conditions. This is one of the reasons that I really appreciate underground construction. There are far more possible locations and conditions where earth-sheltered construction can be used without tearing up the environment than there are for surface housing.

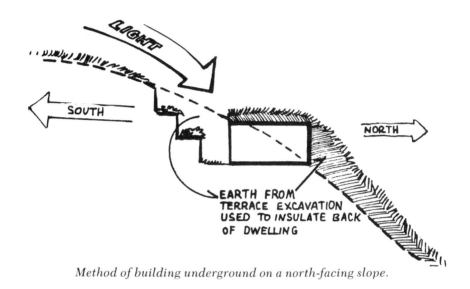

Method of building underground on a north-facing slope.

When using the slope-cut mode for building, I always encourage the excavator to disturb as little of the soil as possible and not to destroy any plant systems surrounding the excavation. These plants not only act as anchors to hold soil in place, but they also provide aspiration to the soil, replenish oxygen and are a vital part of an unnoticed ecological balance. Every year farmers and their city cousins alike have to use more and more chemicals to control insects, pests and unwanted noxious weeds, and encouraging an ecological balance by not destroying plant systems on the excavation makes chemical use less necessary.

Disadvantages of some modes

Each mode has its problems as well as its advantages. The round structure has the greatest bearing strength without excessive reinforcing. Paul Isaacson chose the round shape for that very reason, and the success of the structure in withstanding the expansive pressures over the years, without cracking, has proved his selection. The round shape equalizes all external forces and distributes them over the entire structure, rather than receiving them in isolated locations where a movement may occur.

If a square or rectangle is chosen for the envelope mode, the expansive pressure will have to be dealt with by additional reinforcing of the walls. This can be accomplished by increasing the thickness of the wall towards the bottom in stages. This procedure increases the amount of concrete used but does not increase the reinforcing rods substantially. The forms for this type of wall are more complicated to erect than a vertical form without stages. A uniform vertical wall can be made strong enough by adding more rods and perhaps increasing its overall thickness. (Always consult a good engineer for these figures.) The walls should always be tied to the footings to keep them from moving or "kicking in" at the bottom. No matter what type of structure is chosen, or which mode, I always like a grid type of footing that equalizes and distributes vertical and expansive pressures over the entire base of the structure.

Another problem that exists with all three of the modes, but is more accentuated by the envelope, is that of drainage. Due to the fact that the terrain provides no positive slope or drain, the soil surrounding the envelope is more apt to become saturated during prolonged rain periods. A virtual swamp may be created without a proper drain system to handle it. I recommend the French drain field for all underground structures regardless of type or mode.

The French drain is a simple and effective system. It is a gravel field that surrounds the entire structure and allows water to drain off of and away from the structure without being trapped. This gravel field is tied to a drain tile around the entire footing system that empties into a storm sewer or another absorption area, and it keeps moisture from standing around the structure. Water is very patient and can usually outwait you. Sometimes the footing tiles are channeled into a sump pit, and then the water is pumped up and away from the structure. Here again, I do not like to substitute an electric motor for gravity: Gravity never fails.

Bermed mode: disadvantage

The only real drawback in using the berm mode is acquiring dirt to backfill around the structure. Many times the excavation needed is so small that there will not be enough soil on the site for backfill; and soil will have to be trucked in from another location. This can become costly unless an arrangement can be made with an excavator who needs to get rid of soil from another location where he is working. If the building is located in a rural area, the county road crews may have extra soil they will dump at your location at no cost.

Envelope mode: disadvantages

The envelope does not receive any natural wind for ventilation, so other systems have to be utilized to assure air movement. This is not all bad, since air movement can still be introduced by natural means. Again, Paul Isaacson's double dome served as a plenum that heated from the sun during summer and drew air upward, out of the structure below. Fresh-air intakes brought outside air into a duct system that surrounded the lower wall. This outside air was virtually pulled into the rooms through lower wall vents, then drawn through the rooms, out into the atrium and exhausted up into the dome.

Paul removed several top panels from the dome in the summer to allow the hot air to escape, and thus continuing the air exchange cycle. He told me that this system would exchange air in the summer up to nine times per hour. The late afternoon, the hottest part of the day, was actually the coolest time in the house because of the increased solar chimney effect of the dome. The outside air was cooled and dehumidified as it was pulled through the duct work at the base of the outside wall. (Condensation occurs when there is a ten-degree difference in temperature between a surface and the ambient air temperature.) It is this principle that also added the dehumidification to Paul's home. Any underground home located in a climate that is hot and humid will need a positive air exchange and dehumidification system in the summer.

As I have indicated, most of the problems of the envelope can be overcome without great expense or effort. One last problem that you will just have to live with since there is no remedy, is that of not being able to see out, except upward. As stated earlier, this may even be an advantage in a populated area. If the structure is being built in a scenic area and you want visual contact with that scenery, you probably will not be happy with the envelope mode. In a populated area, I would make my property as inaccessible as possible since there is a tendency for neighborhood children or the curious to walk up and peer down into your atrium or courtyard. If the center is open and not protected by a dome or greenhouse, a railing is needed to keep individuals from falling in for a visit.

An envelope-mode house can provide privacy and abundant recreational opportunities through the use of the courtyard. The security railing can be concealed by plants.

Slope mode: disadvantages

Some of the problems with the slope cut are the same ones encountered with the envelope. Expansive pressure must be dealt with, although not to quite the same extent, since only three walls are below grade. The longer back wall will probably receive the most stress in a rectangular structure. This can be dealt with in similar fashion to the envelope. More reinforcing bars, a staged wall or a thicker wall that is uniform will work with the slope cut. A double wall that serves as an air-movement channel will also add strength to the longer back wall. Reinforcing buttress walls at designated intersections will strengthen a longer span. Deadmen anchors will provide added stability to a wall. If the wall is constructed of cement blocks, a double bond beam will stiffen it against expansive pressure. The wall should always be tied to both the footings and the roof to give it greater shear strength, as well as to help it resist expansive pressure.

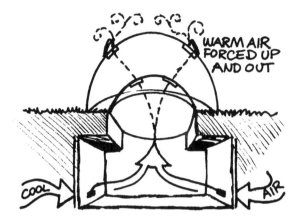

The double dome can help to ventilate and cool an underground structure.

One problem that does not present itself to the other two formats is slope shift. The steeper the grade or slope, the more this possibility exists. Since the soil composition varies and is laid down in layers, the soil tends to slide into any cut or excavation that is created in its face. As the top layer of soil starts to move downhill (perhaps encouraged by a heavy rain) and pushes against the back wall of an underground home, something is likely to give. This shifting creates a

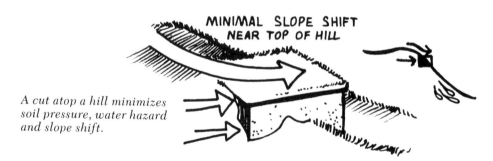

A cut atop a hill minimizes soil pressure, water hazard and slope shift.

"torquing" of the structure. If the structure is located near the base of the slope, it is likely to be pushed out of the face by these forces. When employing the slope-cut mode, I always recommend that the cut be made near the top of the slope in order to minimize slope shift. The problem of underground water (such as a spring or stream) tends to be less near the top of a hill. Water always seeks the lowest level. Runoff increases as it descends during a heavy rain. Combined with slope shift, the runoff creates hydraulic problems for which it is impossible to compensate.

I hope the discussion of the three basic building modes for underground construction will benefit you in evaluating the land you own or are contemplating buying. Unless the ground is located in a swamp or is a vertical face of granite, an underground building will probably work on the site, though the engineering and construction would not be cost-effective.

6
Prairie Cocoon

This chapter will deal with the steps that have to be taken in the construction of any underground building, and specifically my first design, the rectangular Prairie Cocoon. Earth-sheltering is not only the best way to build, but it is also the least forgiving. There are certain steps that have to be taken chronologically to insure that the structure is safe and functions the way it is supposed to function. Surface housing allows many steps and procedures to be bypassed until that particular subcontractor can be on the site. Due to their permanence, many of the procedures and materials involved in earth-sheltering dictate that no steps be missed, otherwise extra expense may be incurred and design changes may be necessary.

Underground building has two major components: terratecture and geotecture. The majority of underground building going on in the United States is terratecture (near-surface construction), and most terratecture structures are cut-and-cover. An excavation is made, a structure that will withstand the loads and stresses of the earth is built in it, and the whole thing is then back-filled.

Geotecture refers to building in deep earth, without breaking the surface soil. The majority of this type of construction is accomplished through either direct earth-tunnelling or a digging-and-shoring process similar to that used in mining. In some cases, where the building takes place in solid rock, explosives are used to blast out a space. Most of the Chinese, Tunisian and Spanish underground dwellings are created under the category of geotecture. Many of these tunnelled spaces have rather conventional façades that utilize adobe, rock or other indigenous materials.

The remainder of this chapter will concentrate on the construction of a rectangular structure that is built from a variety of materials and techniques. This structure is the first underground home that I designed for my family. The main reason that I am using this structure as an example is that it incorporates just about every material that you would consider using in any building, and it will illustrate the methods of applying and fabricating these materials. I will also evaluate the good products and procedures as well as the bad. Since it was my first effort and I had no help or written information to serve as a guide, I made my share of mistakes. Perhaps as I go through the steps of construction, you will get a feel of how your plans will coalesce. It may help you in understanding the frustrations and how to cope with them as well.

There are many shapes and sizes from which to choose when considering the type of structure best suited to your needs. Generally speaking, the rectangle gives the greatest amount of useable space and is the easiest to expand or alter at a later date. These facts, plus the type of topography that I had to build on, influenced my decision to choose this particular design. The following chapters will deal with additional shapes, as well as various techniques and materials with which to construct them. The land that we built on was a ten-acre plot southwest of Lincoln, Nebraska. The land was ideal to build on since it had a south-facing slope and was laid out for visual security. One had to be right on top of the excavation in order to see it. Visual security can be enhanced greatly through the use of plants. Physical security can be effective through the use of plants combined with the terrain (see Chapter 21).

Let me say that if I had it to do over, I would change several things. I will discuss these changes as they would occur during the construction phase. You learn by doing, and no matter how many of these houses or buildings you build, there will always be something you would do differently if you were building them over. I have served as a consultant on many underground buildings, and as I mentioned about designing airplanes, they have to fly before you find the bugs. Underground houses are no exception to this rule. Once the building is completed, the real feel for how it is flying begins. Also, like airplanes, underground houses can be based on a single design, but each one will be different in its feel and performance to some extent. This is because the materials may vary slightly, and the earth has a different makeup and personality wherever you build. People tend to forget that the soil is a living and breathing entity, and if you let it talk to you, your building will be much more successful.

The soil

The soil at the building site of my first underground home was an expansive clay which required some careful planning with regard to footings and bearing capacity. A rectangular shape was chosen for several reasons. I did not want a deep excavation to accommodate a round or square structure. A frontal greenhouse was an integral part of my plan and required width in order to achieve maximum efficiency as a heat collector and a light dispenser. The precast roof panels I had chosen would not permit the use of skylights; therefore, the depth had to be limited in order for the light from the greenhouse to penetrate to the rear of the structure. The rectangle not only provided the greatest amount of useable space, but it also helped solve the ventilation and air distribution problems. The available labor skills and materials also lent themselves to this shape more readily.

Building on any expansive soil requires that the bearing weight of the structure be spread over as large an area as possible. This is the job of the footings. A great many surface structures have poor footings, resulting in settling and cracking. If this occurs in an underground house, the results could be fatal. The loads that are imposed are great, and one must account for them. I worked out a grid system that would distribute the weight over the entire floor area instead of at the perimeter and specific bearing points.

Footing design

The footings are the most critical part of the structure and should not be compromised. There are a number of areas where compromise will save dollars, but the footings are definitely not one of those areas. Generally speaking, the width of the footing is more important than the depth; the actual shape of the footing can be a help or a hinderance. A footing with a wider base will present the greatest surface to the underlying soil and thus the greatest amount of resistance.

The footing can taper towards the top until it is the same width as the wall that rests upon it, helping to prevent pooling ground water at the base of the wall.

The Prairie Cocoon shown above is one which I built near Martell, Nebraska. It is an excellent example of how a working greenhouse can be utilized as a source of heat, food, oxygen and recreation. The use of recycled railroad ties for earth retainment is both economic and aesthetically pleasing.

The overall structure of this underground house is based on the Prairie Cocoon. Everything in this home is accessible by wheelchair, including a special bathroom with a drive-in shower. Necessary conveniences are operated electronically by touch control, and the lights turn on and off through the use of heat sensors.

Most footings poured by many contractors have vertical sides, since this is the quickest method. A backhoe makes a trench, and the trench is then poured. This is detrimental to the longevity of the structure and is more costly in concrete to

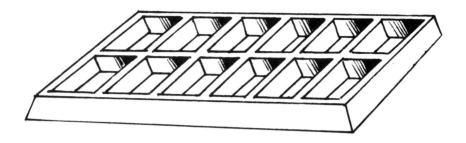

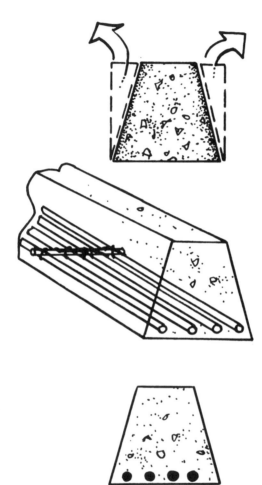

Above: Grid-type footing which spreads the weight of the structure over a larger base is used when building on expansive or poor bearing soils. A bell-shaped footing, left, offers the best bearing strength with the least amount of concrete.

the owner who pays for the materials. Flexible plastic drain tile placed next to a footing that tapers up to the wall will keep water from collecting at the base of the wall and from finding its way inside. Tapering the footing not only saves concrete, but is also just as strong as one with vertical sides since the upper corners do not add to the shear strength of the footing. This is something that I did not know when I built the structure and, consequently, wound up with moisture problems at the base of the walls. I also wound up buying a lot of unneeded concrete.

The steel reinforcing rods that are placed at the bottom of the footings are very important to the strength of the footing. This steel actually ties the concrete, making the footing into a beam that will have great shear strength. I placed four reinforcing rods lengthwise in the bottom of my footings and tied them with short crosspieces every six feet. All of the junctions with other interior footings in the grid system also were reinforced in the same manner, and the intersections were all reinforced with extra rod.

The footings need to rest on a firm base. The clay that I had to build on was like concrete on the surface during dry weather, but became very spongy during wet seasons. I did several things to help compensate for this problem. First of all, it is best to disturb the soil as little as possible during excavation. An excavator with a good reputation is far better in this case than a friend or neighbor who is willing to donate a tractor and loader, etc. This is another area where shopping pays off. Among good excavators, the price for your job may vary considerably. Do not buy the cheapest service just because it is cheap.

There are legitimate reasons for sizeable cost differences; just be sure they are legitimate. One reason for less cost from a good excavator may be the fact that he is in the area working on other jobs and will not have to transport equipment and personnel very far. Another reason may be that one or two contracted jobs may have been delayed, and he needs to keep his crews and equipment busy. Once again, check credentials and ask previous customers about the quality of the excavator's work.

A good guideline is to see how close the excavator can make the cut to the required specifications. If the hole is overdug and fill is required, I would be cautious. If I would have had to add fill soil to level the excavation for building, I would have been adding to the possible settling problems already inherent within the soil. Make the cut as perfect as possible, and the soil then will not need additional compaction.

Excavation equipment

Excavation equipment comes in two basic types. One type is the Caterpillar tractor (Cat), which has a digging bucket. The other type is the backhoe, which is usually mounted on a rubber-tired vehicle, and pivots left and right as it lifts the soil up and out of the excavation. Both systems have their respective merits and drawbacks. I contracted with a firm that used the crawler with steel tracks and scoop. The main advantage here is speed and the ability to track the dirt to a more desirable location for future backfill purposes. The disadvantage of the tracked vehicle is that it tends to crush more plants in and around the excavation, and it packs soil that does not need compaction. I chose this system because of the need to track the removed soil up and around the excavation.

It was easy for the Cat to round out my sewage lagoon at the same time and track this extra soil to the house-site for additional berming. I was building on ground that needed to be reconditioned and revegetated with native plants. The tracking and soil abuse was confined to the smallest area possible in order to control ecological damage.

The backhoe's advantage is that it is easier on the environment and tends to work from a fixed position while lifting the dirt out of the cut. If you are building in a confined area or in an area that has a good ecological balance, you might do well to consider choosing the backhoe. If you are a self-builder with a good back, hand-digging tools, a wheelbarrow and lots of time, I would consider digging the

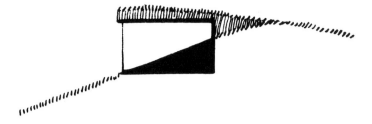

Shallow slope cut near the crest of a hill provides the best soil stress conditions for the Prairie Cocoon.

required hole by hand. This guarantees the least amount of damage to the environment, and you can dig as precisely as the digging stakes indicate. My next underground home will be built entirely from indigenous materials, the topography will be utilized as it exists and any soil removal will be done by hand. For many people, this system is impractical, but it is the ideal approach.

Excavation process

All of this discussion about footings presumes that your structure will need them. The use of heavy materials requires a footing or a solid base (such as rock) on which to build. Poured concrete, cement block, brick, prefab concrete units, and stone all fall in this category. My home used just about all of these materials. A post-and-beam design would not require footings and in some cases is preferred for that very reason. The post-and-beam method fits into a variety of terrain conditions where it would be hard to adapt footings.

Continuing with the step-by-step construction, the excavation was located at the crown of the slope. The cut was very shallow and was only six feet deep at the northwest corner of the back wall and two-and-a-half feet deep at the northeast corner. The cut was level and ended at the front edge of the slope. If the excavation is being made with a Cat, be sure that it is dug large enough to accommodate a backhoe at the perimeter of the proposed structure. I did not do this and had to dig the back and side wall footings by hand. The temperature was very hot, and the clay was solid. Immediately after the excavation was dug, it rained for several days and filled the footings, washing dirt into them, as well as the excavation.

A decision to build a bermed structure was made for several reasons. The land did not slope enough to do a wall cut and the water table was very high. The digging of the sewage lagoon provided additional soil that would create a berm large enough to blend with the rest of the terrain. The length of the back wall and the fact that I chose to build with fifteen-inch cement blocks dictated that I minimize expansive and shear pressure. The berm allowed the least amount of stress on the structure while maximizing thermal efficiency.

Walls: preliminary information

Cement block was chosen as the building material for the walls for two reasons. My neighbor was a small contractor who specialized in block-laying, and due to his proximity to me, the cost of laying the cement block was reduced since the amount of his time was minimal. A nearby block plant was most happy to have a customer close at hand and consequently, the price was satisfactory. These costs, when compared to the cost of erecting concrete forms, the cost of pouring and the per-cubic-yard cost of ready mixed cement, gave a healthy margin in favor of laid-up walls. The three outside walls were laid up in conventional fashion with mortar. The cores of the blocks were filled with cement on six-foot centers. These filled cores also contained four reinforcing bars, two per core. These vertical bars in the cores made columns that gave added shear strength to the walls. The long rear wall was bolstered through the use of buttress walls on twelve-foot centers. These walls served as interior room-dividing walls. The buttress walls and the filled cores gave the walls adequate shear strength, but to protect them from cracking vertically between these supports, a bond beam was created as the top course of blocks.

Explanatory terms

Definitions of these terms may help at this point to understand the processes involved. *Cores* refer to the holes in the cement blocks. A wall laid up in standard overlapping pattern leaves these holes in alignment so as to form vertical tubes. *Shear* is a term that applies to any material that has to withstand pressure or a load. The stress point at which this material breaks or fails is referred to as shear. There is both vertical and horizontal shear potential on a wall that is backfilled. The soil pushes against the wall so that it would break inward on a horizontal line without vertical support. The filled cores and buttress walls prevent this. The expansive pressure would cause the wall to bow horizontally as well as vertically. The wall would break inward like a double door opening without horizontal support. The bond beam prevents this. A *bond beam* is created in a masonry wall through the use of a special block. This block is "U"-shaped with no top, just sides and bottom. When these blocks are laid end to end, they form a *trough*. This trough is then poured full of concrete after several bars of reinforcing steel rod have been placed in the bottom; thus a beam is created with steel reinforcing. This beam keeps the double doors closed.

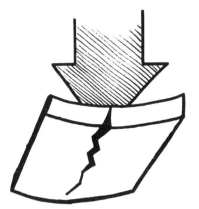

The bond beam at the top of the wall presents this type of shear where the wall bows in from soil pressure.

Filled block cores that form posts within the wall prevent the type of shear shown at right, which is caused by soil pressure.

This system worked very well for me; two years later there were no cracks. Other ways to help relieve soil pressure against exterior walls will be explored in following chapters.

Surface-bonding

One thing I would do differently with these walls if I were to do them over is to surface-bond them instead of using standard mortar joints. This system helps to waterproof the walls, but the most important thing it does is to make the walls four to six times as strong as a mortar joint wall.

Let me explain what surface-bonding is and how it works. Instead of bonding the blocks to each other with cement in the joints, a bonding compound is applied to the exterior and interior of the wall, covering all joints and block surfaces. The blocks are dry-stacked, and no mortar is used (mortar does not bond the blocks anyway). Most cracks found in block walls usually follow the mortar joints. Mortar is used mainly as a levelling agent. The compound that is used in surface-bonding is primarily cement and fibreglass strands that are about one inch long. It is these strong fiberglass strands that give this agent its strength. Millions of these strands run in all directions, overlapping joints and each other to form an unbreakable matt on both sides of the wall.

Some people have asked me if it wouldn't be even better to make a standard mortar joint wall and then surface-bond it. I believe that this would actually weaken the wall instead of strengthen it. The reason for this is that the joints add space between the blocks that would not receive as many fiberglass strands completely across these spaces, thus bonding the blocks. When the blocks are

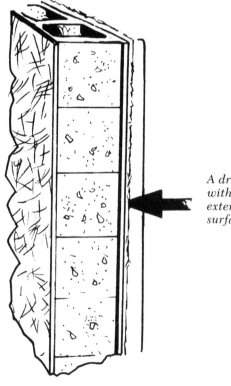

A dry-stacked block wall with an interior and exterior bonding cement surface, left.

dry-stacked, the space is minimal and almost all of the strands bond completely across the joints onto the blocks. During the underground seminars at *Mother Earth News*, we dropped and tipped several bonded wall sections onto a concrete floor. Some of the blocks cracked, but not one joint failed.

These bonding agents are made by several manufacturers and all have about the same ingredients; check with your local building supplier. One note of caution: If your soil has a high alkali content, check to see that the bonding agent being recommended will not be etched or eroded. The bonding agents can be sprayed or trowelled. It is my recommendation that they be trowelled onto a uniform thickness. It is this uniformity that gives the wall strength. Dry-stacking makes the amateur look like a professional and allows the amateur to correct his mistakes before they become permanent.

The use of surface-bonding combined with filled cores and a double bond beam would produce a wall that would stand without buttress walls. This would allow the possibility of a clear-span structure that could be altered at a later date should a change in space use be desired (more on this later in the chapter). Surface-bonding, when properly combined with other support systems, will be as good as a poured wall, and in some cases, such as mine, more economical. Here again, labor, materials, transportation and skills have to be considered.

Waterproofing and drainage

Once the footings are completed and the walls are in place with all of the reinforcing components, it is time to take care of waterproofing and drainage. Placing drain tile around the footings is vital.

I dug an eight-inch trench next to the footings, and the top of the trench was level with the top of the footing. This allowed the runoff from the drain field above to reach the tile and get away from the structure. The eight-inch-deep trench was lined with about two inches of coarse river gravel. The wire-supported, flexible drain hose was placed on the two inches of gravel, and the hose was then covered with the same river gravel until the area was level with the top of the footings. The three-foot-wide footings trapped water next to the walls. The water did not reach the drain tile, and it caused moisture problems with the walls. If I had cast the tapered footings, this water would have run immediately into the drain-tile trench and away from the walls.

The drain tile ran completely around the structure and exited at both ends of the front retainment walls. This allowed the water to flow away from the building, down the slope. After the drain tile was in place, the need to waterproof and drain the wall areas was my next concern. At the time I built my first underground home, there was very little material available that would apply directly to underground buildings that would serve as a waterproofing agent. Most of the materials in existence had a bitumen base and, in my opinion, were next to useless as a permanent waterproofing agent. This is certainly an area that today is not lacking in materials. Trying to decipher all of the claims is next to impossible. After talking to various suppliers and builders, I bought a cold asphalt to trowel on the outside of the walls. This was then covered with a sheet of six-mil plastic. I would have probably been just as well off if I had not used the cold asphalt. Asphalt tends to dry and crack. In some cases it can even contribute to the moisture problems by trapping moisture in the asphalt cracks next to the wall. The plastic sheeting gave the wall its protection. It is my opinion that a simple raincoat for the walls is all that is needed if a drain field surrounds the entire building.

All drainage water should be channelled away from the structure so that water does not pool.

There are many good waterproofing products on the market, and the number is increasing. These agents fall into various categories: bentonite, sheet or roll membranes, and liquid sealers. Bentonite is a dry powder that can be irrigated and sprayed on, or it can be bought in sandwich panels that are fastened to the structure, and then irrigated. Since bentonite is a refined clay, it will always stay expanded, thus keeping out additional water molecules. This means that there

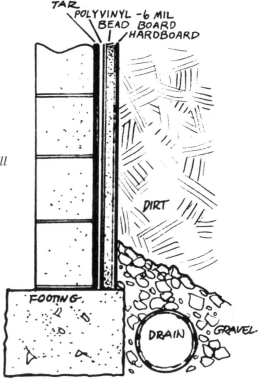

Standard mortar joint block wall illustrates a poor method of waterproofing, since the footing will catch moisture.

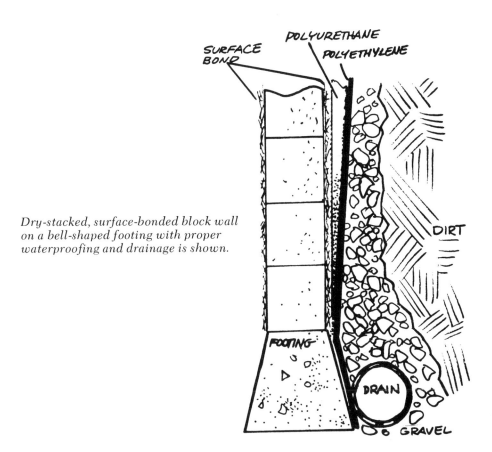

Dry-stacked, surface-bonded block wall on a bell-shaped footing with proper waterproofing and drainage is shown.

will always be a damp surface against the outside of your walls. If I were using bentonite, I would still raincoat my walls with plastic sheeting. Concrete is very porous, and without this raincoat, the concrete would stay damp on the inside walls. Bentonite, when properly used, is probably the most permanent form of water protection available.

The next category of waterproofing agents is comprised of sheet or roll membranes, which are plentiful. Bituthene, Gates N-3 Neoprene, Sureseal, a butyl sheet by Carlisle Tire and Rubber Company and others make up this category. My main reservation regarding sheet membranes is that they all tend to be expensive. Some large commercial underground buildings may be required to go this route in addition to using liquid sealer in order to satisfy insurance company requirements. Liquid sealers comprise the third category. Any one of the plastics or butyls on the market do a good job. Here again, you have to decide what is cost-effective for you.

If I were doing my home over again, I would do several things differently in regard to waterproofing and insulation. I used two-inch-thick sheets of beadboard for insulation, which was placed on top of the plastic sheeting used for waterproofing. With what I have learned in the intervening time, I would not use the cold asphalt, but would use double sheets of polyethylene against the surface-bonded and polyurethane-sprayed walls. I would have polyurethane sprayed on the entire structure. For our Nebraska climate, I would spray three inches on the roof and two inches on the top of the outside walls, reducing its thickness to about a half-inch at the bottom of the walls. (Note: In very cold climates, footings and floors should also be insulated so as to avoid bleeding.)

A wall cross section would start with fifteen-inch-thick cement blocks that are sprayed with polyurethane, two sheets of polyethylene, four inches of coarse river gravel and a fiberglass cloth matting (to keep the soil from plugging the gravel field). All would then be backfilled with soil. This sounds elaborate, but it is a surefire system and less costly than other systems of waterproofing and insulating.

Earth-covered roofs

After completing the footings, walls and drain fields, the next thing on the agenda before backfilling the walls is the roof. I had several roof systems in mind when I first began thinking about the Prairie Cocoon. Several builders tried to talk me into berming the walls and then putting on a conventional roof. This is an energy-efficient system, but it lacks some of the things I wanted to achieve through an earth-covered roof. The grass or plants on an earth-covered roof provide additional oxygen to the environment. An earth-covered roof tends to spread and dampen the effects of weather on the structure: Hail, tornados or other severe weather conditions usually do little damage to an earth-covered roof. Such a roof is fireproof, which is comforting when a chimney fire occurs and molten material is spewed upon the roof. The soil roof blends with the rest of the terrain and is easier on the environment from an aesthetic point of view while also providing a habitat for small animals and food for birds.

Roof construction

A roof made of precast concrete panels is sometimes preferable to a poured-in-place roof. I chose Flexicore panels for the roof system. Flexicore is a franchised name for these panels, and units of similar design may not necessarily

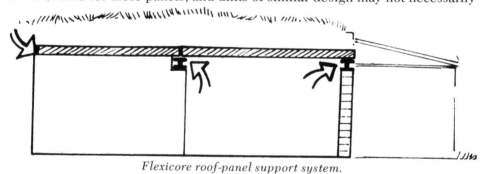

Flexicore roof-panel support system.

have this name. These panels come in several thicknesses and strengths. I chose the eight-inch-thick panels and secured them with six tension cables. The cross section shows how these units look from the end. They are cast in sixty-foot lengths out of a heavy limestone aggregate concrete. Six steel cables are drawn up under five thousand pounds of tension. The cables are anchored at both ends of the form and are released after the form has been poured and the concrete has set.

As the cables contract, they stress the unit the same way you would stress a number of books by standing them on end and then pressing them inward with both hands. By doing this you create a book beam the same way the cables create a roof beam. These units were cut into fifteen-foot lengths to fit my specifications. The units rested on the outer back wall, a front girder supported by columns on seven-foot centers, and a center girder. The units were transported by three semi-trailers and lifted into place by a mobile crane. The entire two-thousand-square-foot roof was completed in two hours. After the units were in place, they were grouted with thin concrete which tied them together with the one-inch rebar in their keyways. The rebar in the keyways had tied the units together over the center support beam.

The entire roof was poured with a coat of concrete. This cap was four inches thick at the front of the roof and tapered to two inches at the back edge. This cap helps to transfer the imposed earth load (and any other live loads) over the entire roof. Since the individual Flexicore units are similar to planks laid side by side, each plank or unit would not transfer any load that it would receive without its concrete cap, thus possibly bearing too much disproportionate weight. The drop of two inches from front to back of the roof cap provides drainage without puddling.

The tension cables in the units cause the units to bow slightly upward when in place. This is intentional, so that when the units are loaded they will flatten out, but not deflect. The casting of these units has to be done with great care to insure uniformity when they are in place.

Walls and their relationship to the roof

The walls receive part of their support and strength by being tied to both the footings and the roof. They are tied to the footings by a reinforcing rod that is left protruding from the footings. These bars occupy the cores of the wall blocks which are poured with concrete. Anchoring the wall to the roof is a little more complicated.

There are several forces at play that want to destroy the walls. The soil pressure wants to bow the walls horizontally and vertically and also acts like a lever to tilt them inward. Expansive pressure tends to be greatest at the base of a wall, but without being tied to the roof, the wall would be levered inward at its top by the heaving soil. The soil nearest the surface is more apt to be affected by freezing and thawing, in addition to greater expansion from moisture absorption.

Placing the roof panels just to the center of the side and back walls helps to stabilize the walls against the levering effect of the soil (see illustration). Before the walls could lever inward, they would have to raise the entire roof. By drilling into the top course (the bond beam), half-inch rods can be inserted in the holes on two-foot centers. These holes are drilled at the edge of the roof panels where they rest at the center of the bond beams. The rods that are inserted into the holes are then bent over the roof units. This all occurs before the cap is poured on the roof units. The forms that keep the capped cement from running over the walls also create a trough around the edge of the roof units. When the cap is poured, the trough is filled and the walls are tied to the roof and its cap. This concrete bond and the placement of the units on the wall assure that the walls will never lever in at the top.

Twin Tees vs. Flexicore

Flexicore worked very well for the roof, but if I were doing it over I would use Twin Tees in place of the Flexicore panels. The main drawback of Flexicore is that light shafts or skylights cannot be inserted into it. The Flexicore units are only two feet in width, and any openings that would be cut in them would either damage one of the hollow cores or would cut supporting cables. Twin Tees, on the other hand, are created with a flat deck that is three feet wide between the supporting legs of the platform. This width allows light shafts to be cut three feet wide and up to twelve feet in length. Skylight boxes can be cast that will set down into these openings, resting on a lip that contains a sealing gasket. The ability to clear a span of up to fifty feet (even more with specially cast units), is also another reason that I would use Twin Tees. The cost of the Twin Tees is considerably higher than that of Flexicore; however, the savings incurred by not having to build the buttress walls should about equal the additional cost of the Twin Tees.

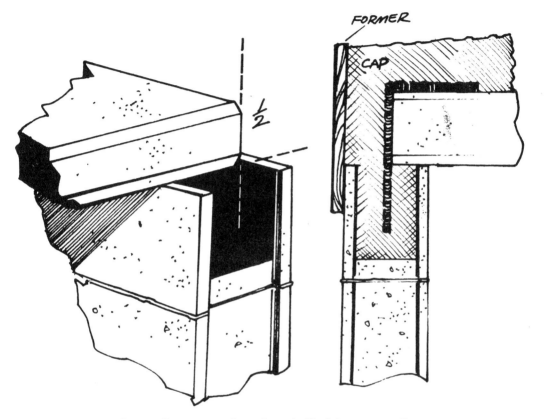

Placing Flexicore roof panels on half of the outer walls gives the walls cantilevering strength.

Like the Flexicore roof, Twin Tees have to have a cap of concrete poured to distribute load, but the wall-anchoring procedure would be different. Elimination of buttress walls would dictate the use of an additional bond beam about midway up the wall. The bond beam at the top of the wall should remain the same. The Tees can rest on the outer back wall's bond beam and overhang the front beam-and-column system to provide the sun shade in the summer. Since the decks of the Tees will be a couple of feet higher than the outside walls, the ends and sides will have to be filled with extra courses of blocks and some concrete. The freedom of being able to change the internal space as the family grows and changes in its space requirements is important. This option makes any additional cost reasonable, and this spacial flexibility serves as a positive factor when you choose to sell your home.

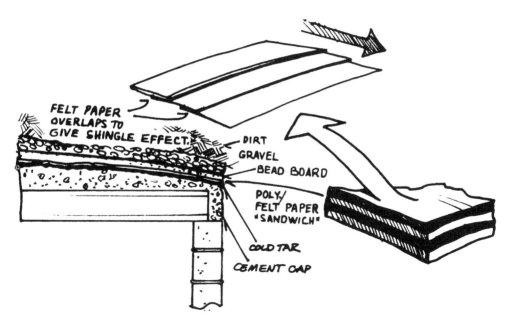

Cutaway shows the roof strata.

Waterproofing the roof

It is extremely critical that the roof be properly waterproofed. Waterproofing should be looked at from the perspective of getting the water away from the roof as quickly as possible. The same waterproofing agents that work on the walls will work on the roof. I used the cold asphalt that I used on the walls and then covered the entire roof with a specialized roll roofing. This roll roofing was a nonbiodegradable sandwich of treated felt paper and two-mil sheets of polyethylene. The bottom layer was felt paper; the next layer, two mil polyethylene; the third layer, felt paper; and the fourth layer on top was polyethylene. These various layers were all laminated together with a compound that acted as an adhesive. The rolls were four feet wide and forty feet in length. Since the roof sloped towards the rear, a standard overlapping pattern was used, starting at the back and working towards the front. A butyl strip was rolled down over the overlaps to help seal against possible infiltration during heavy rain or prolonged wet periods.

Sheets of beadboard insulation were placed on top of the roofing, then two inches of coarse river gravel was hand-shovelled on the boards. A one-foot layer of black soil topped the fill. The top layer of black soil was seeded with Kentucky Fescue grass, a broad-bladed grass that does well in hot, dry conditions. It has a tremendous root system which prevents erosion. I created mini-terraces, using railroad ties to help hold the soil until the grass took over.

Just as I would treat the walls differently, I also would treat the roof differently if I were building the structure over. After the concrete cap was poured on the Tees, I would spray three inches of polyurethane over the concrete. I would then protect the urethene by covering it with two sheets of polyethylene, just as suggested for the walls. The gravel and fill would be the same as that on the original structure.

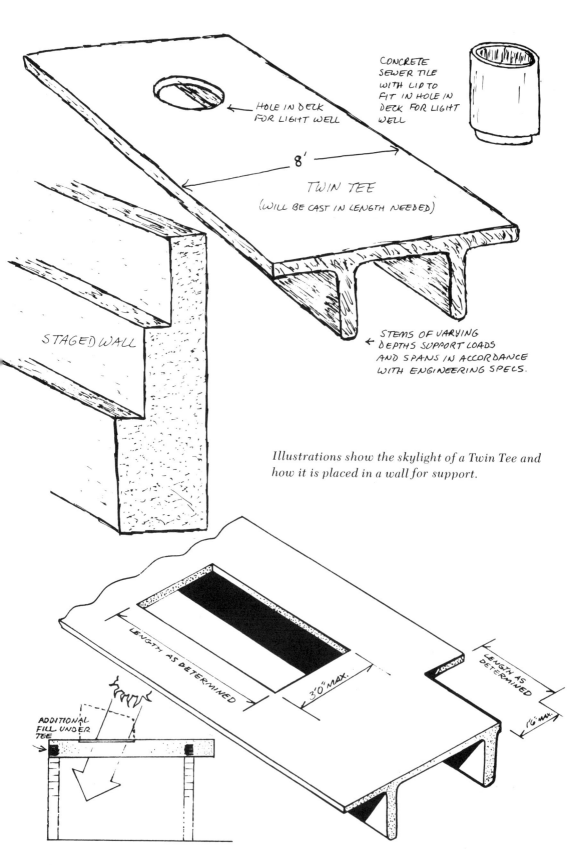

Illustrations show the skylight of a Twin Tee and how it is placed in a wall for support.

Utilities, wiring and plumbing chases

Before any backfill is accomplished, be sure that all of the utilities, wiring and plumbing chases are completed. This is one part of the step-by-step process that I thought I had taken care of, since the gas, water and electricity had been run in their chases, with plenty of access provided. I had even provided for special sewer cleanouts to avoid the possibility of having to break up concrete at a later date. It was not until we were ready to move in and I inquired about establishing phone service that I realized there were no buried phone lines. There were no phone lines above ground, either. Finally, I talked the phone company into using their slitter that buries cable directly in the ground. They buried the wire up to the edge of the garage. The wire was hidden from view by a trim board up to the point where it entered the garage. Once inside the garage, I ran the wire down the center support beam. This beam was boxed in up to the point where it entered the garage. This provided an aerial chase. The wire was then run down the hollow wall that divided the kitchen from the main bathroom. The phone was then connected to an outlet that was accessible in the kitchen and living room area. All of this would not have been necessary if I had thought of *every* step.

Constructing the greenhouse

The greenhouse, which I considered the heart of the house, was my next project. The footings for the greenhouse were dug and poured at the same time that the footings for the main house were prepared. The greenhouse footings did not have to be as wide as those of the main structure because the greenhouse was extruded and a light-weight structure with no earth load imposed. Footings for the greenhouse did have to extend below frostline to prevent the soil inside the greenhouse from freezing. The footings were dug fifty-six inches deep. A twelve-inch-thick base was poured with eight-inch cement blocks laid to ground level. The blocks were topped with a concrete cap that contained stud bolts in order that the cap would be fastened to the base plate of the front wall of the greenhouse.

Finding a way to attach an extruded building to one covered with earth and made of dissimilar materials was an interesting task. The greenhouse was a frame structure with a metal roof that had to attach to a steel-and-Flexicore façade. It was decided to drill down through the cement cap into the cores of the Flexicore at the front edge. Threaded, one-half-inch rod was bent in ninety-degree angles and run through cedar 6 by 6s on two-foot centers to match the holes drilled in the cores. The cedar 6 by 6s were lifted up, and the threaded rod hooks were inserted into the holes of the cores. The 6 by 6s were then aligned, and the bolts were drawn up tight. The 6 by 6s became the face plate to which the rafter hangers and rafters for the greenhouse were attached. Two by fours were glued to the front "I" beam that supported the Flexicore units. These 2 by 4s were glued against the stem of the steel beam as well as the bottom flange. The 2 by 4s became the face plate for the lower greenhouse rafters.

The entire front frame of the greenhouse was constructed on the ground in front of the footings. It was then tilted up onto the footing and the rafters were nailed into position to support it. The end framing for the greenhouse was nailed into position and the metal roof was then applied. This lightweight overlapping metal roofing was chosen for two reasons: It was the most economic and long-lasting roofing available, and the green enamel paint blended with the grass roof of the main structure. Due to the shallow pitch of the greenhouse roof, the metal would drain most rain quickly and let heavy snow loads slide off without building up. These metal panels were nailed to two-by-two stringers that ran across the rafters on two-foot centers. This system proved very strong and would support over two hundrd pounds per square foot.

The only problem encountered while joining the structures was attaching the flashing to prevent water infiltration between them. My first mistake was to make the flashing too narrow. My second was to use cold asphalt as a sealer. As a result, I had a leak the following spring after the ground thawed. After two or three attempts to stop the leak, I decided to take up the flashing and do it right. I used butyl rubber in liquid form as a sealer this time and solved the problem.

Stops for the glass were made of one-by-four facing boards that finished the four-by-four framework of the greenhouse. Four-by-seven-foot custom-assembled thermopane was installed to enclose the front. The three-sided glass enclosure added a dimension to the house that even our imagination had lacked. I insulated the roof of the greenhouse with six inches of rockwool and finished the inside ceiling with drywall. We spent more time in the greenhouse than in any other part of the structure. My daughters used it to keep their tans all winter, we grew food and raised fish in a four-by-six-foot fish pond. The greenhouse turned the house into a nearly self-sufficient, closed environment. It created food, oxygen, heat, light and recreational room. I laid loose brick directly on the dirt floor of the greenhouse and simply removed the bricks in the areas that needed planting. This flexibility in space use was the best part of the design. The extruded position of the greenhouse also allowed one to be able to see out to the east and west, as well as south. Light distribution from the greenhouse was excellent.

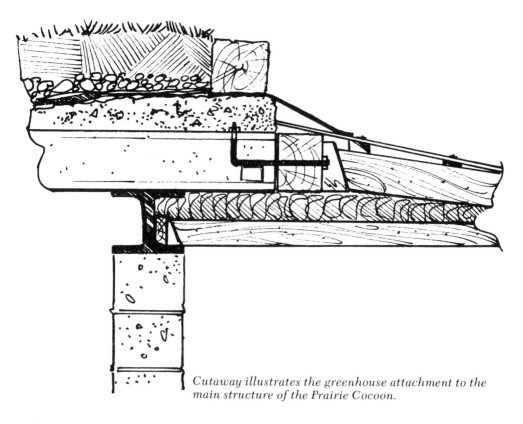

Cutaway illustrates the greenhouse attachment to the main structure of the Prairie Cocoon.

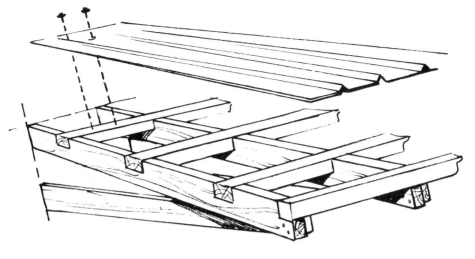

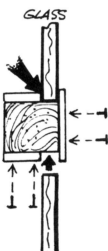

Rafter support system for the metal roof, above. Left: Simple framing and stops for the greenhouse made glass installation quick and easy.

Light distribution

As I stated earlier, skylights in the rear of the house would have balanced the light. Light balance is as important as the amount of light. The greenhouse was so bright that the house seemed dark once you turned and faced the back part of the house. After your eyes grew accustomed to the lower light level of the rear part of the structure, it seemed as full of light as the average room in a surface structure. It was the vast difference in light intensity that made one part seem dark and the other nearly blinding at times. This effect was intensified by the accumulation of snow around the greenhouse, since the snow acted as a reflector.

Another thing we found out about light intensity was that not all plants enjoy direct or bright light. We had to shift plants in the greenhouse according to season. The winter months allowed the sun to penetrate directly into the greenhouse because of its low angle. The summer gave us the opportunity to bring the light-sensitive plants towards the front of the greenhouse due to the sun's nearly vertical angle. A further note regarding sun angles: I calculated that a two-foot overhang on the greenhouse would be adequate to shade the glass through the hottest summer months. It was not. By the end of August, the sun had dropped low enough to strike the bottom third of the glass, and by the middle of September the angle had dropped until almost half of the glass was receiving radiation. The heat gain was too much for the earth's cooling effect to compensate, so I used aluminum foil halfway up the glass until the season changed enough to make us appreciate this heat gain.

A permanent system that allowed light into the greenhouse with direct solar gain was worked out. It consisted of a series of shutters with movable slats that deflected the radiation from the glass while allowing sufficient light. The shutters also protected the glass from hail and high winds. These side benefits were a result of poor original planning, and it would have been better to avoid the expense by better preplanning.

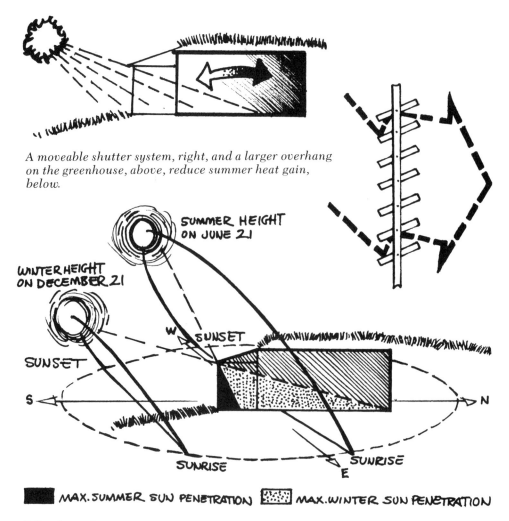

A moveable shutter system, right, and a larger overhang on the greenhouse, above, reduce summer heat gain, below.

The interior

One of the areas where cost-cutting is a matter of preference in materials and will not compromise the integrity of the building is the interior finish. I realized I had to start making some real compromises on the interior finish when my construction costs kept rising and I had secured a third loan for the construction. Much of the interior was finished with drywall to give the house a feeling of convention since, at the time, the concept of an underground home was considered radical. I felt this would help with resale, should I ever decide to sell the house. I held with this idea fairly well, but I did use rough exterior cedar panelling in some rooms as contrast walls in order to trim costs. I gave up the idea of putting up a separate ceiling to cover the Flexicore units, and I used drywall compound and textured the Flexicore instead.

Flooring

The floor was another one of those unplanned happenings that worked out but cost twice what it should have. Originally, I had planned on just pouring four inches of concrete and then adding a specially selected aggregate of washed river rock. The river rock was in turn supposed to be ground off to give the appearance of marble. I had seen a floor like this before, and it was absolutely beautiful. Well, my cement finisher had neglected to tell me that he had never run aggragate in a pour before, and the result was disastrous. He tried to pour almost two thousand square feet of floor at one time, then go back and add the aggragate. Of course, none of the aggragate adhered to the concrete. I had to scoop it off the concrete, which made the concrete look as though it had a bad case of smallpox. Fortunately, I had a good foreman who was also my chief carpenter. He helped me work out a system to not only cover the mess up, but also make the floor attractive and easier on the feet, legs and back. (It is worthy to note that a good foreman can be one of your biggest assets, especially if he has prior building experience.)

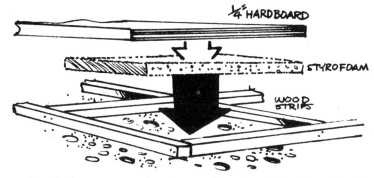

The subfloor is covered with Styrofoam board, hardboard and brick.

The new system my foreman recommended called for one-by-two nailing strips to be shot down on the concrete every four feet. Half-inch Styrofoam sheets were placed between the nailstrips. The whole system was then covered by quarter-inch hardboard. Some rooms (such as bedrooms) had carpet placed over the hardboard, while the bathrooms were covered with linoleum. The living room and kitchen were covered with one-and-a-half-inch-thick floor brick that was laid in geometric patterns. The kitchen brick was grouted and sealed to protect it from water and cooking accidents. It also made the floor easier to clean. The living room was not grouted, and the bricks were allowed to "float." At first people thought I was crazy, but my theory proved out. The slight give in the system really felt good on the body. As I said, the floor system really worked great, but it cost double what it should have cost. If I had it to do over now, I would not put any cement down at all. Cement is too unforgiving, and if one stands or walks on it for an extended period of time, it can cause discomfort in the back and legs.

The built-in cabinets and open shelves of the kitchen, above, are constructed of one-inch cedar. The "L" configuration left little counter space, so a Danish butcher block was made out of 2 by 12s. At left is a view of the main bathroom. Included in the room are a long vanity, double sink, tub/shower, stool, storage space and power ventilation.

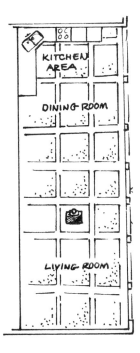

One type of floor plan for an earth-sheltered home.

Air management

Probably one of the most neglected conditions in an earth-sheltered home or structure is the need for positive and controlled fresh air input. Due to the nature of earth-sheltered structures, the internal air pollution often exceeds that of the outside world. In the cold winter months, when a furnace or fireplace is in use, the infiltration around and through windows, doors and walls will produce ample supplies of fresh air when combined with the inefficiency of the furnace. Often, the fireplace (if flue is open) is a net heat liability. This same system in summer, relying entirely on infiltration, is greatly lacking, and the structure usually has a lot of stagnant air pools. If this stagnant condition exists in a humid environment, one will experience hot and humid stagnant air, which is uncomfortable and unhealthy.

One method of controlling this adverse condition is to employ a fan system. This may be accomplished by using the furnace fan itself as an air handler, or by installing a fan (such as an attic fan) in a conventional home. Unwanted hot air can be taken out of the structure through infiltration points, or it can be released directly to the outside. Fresh air can be introduced into the interior through ducts or to the furnace fan.

Perhaps the best answer to the air management problem is the use of cool pipes (earth pipes, cool tubes, etc.). This is a system of pipes or air ducts that are buried in the ground so as to temper the air coming into a structure. The air which enters through the cool pipe is very close to the temperature of the earth. On hot days, the hot air comes in contact with cool pipes of considerably cooler temperatures. The ground has the ability to absorb tremendous amounts of heat and still remain relatively the same temperature. Therefore, hot and humid air is lowered in temperature, and the amount of moisture in the air is also greatly reduced (by as much as 60 percent).

Because of the amount of moisture removed from the air in the cool pipes, a good way of handling the evacuation of the water from the cool pipes is to use them in connection with French drains. This not only keeps the French drain clean and fresh, but allows for condensation moisture to escape and be drained away. This system, however, is not a new innovation; perhaps the oldest one in use today is the cool-pipe system running under and into Mt. Vernon, Virginia. There has been much experimentation along these lines, and as of this writing no one has been able to generate "the" formula for designing these air systems.

The smaller the diameter of the pipe, the larger the ratio of surface area to the volume of air moving through a pipe. On the other hand, sufficient volume to provide for habitation needs is of paramount importance. In either case, the straighter and smoother the pipe, the better. Corrugated pipe may be readily available and inexpensive, but there is a potential for standing water in the "ribs." This water will then produce fungal growth as well as odor.

The cool pipe or earth pipe works equally well in the winter to warm the air to the furnace or heating system. Outside air of zero degrees Fahrenheit is drawn through the earth and is in turn heated or tempered to around 55 degrees Fahrenheit (or more in some cases) prior to being introduced into the heating system. This acts as an "air preheat."

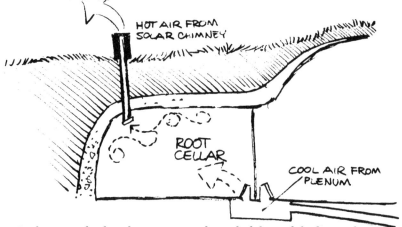

Soil pipe and solar chimney provide cool, dehumidified outside air, creating top and bottom ventilation.

In conjunction with the cool pipes, it is advisable to incorporate the use of a "chimney effect," which may be of almost any design as long as the function is to extract air by passive means. A solar chimney works best on sunny days to extract hot stale air. In principal, a solar chimney has a heat collection section of dark material inside a transparent box. As the sun heats up the air inside, natural convection lifts out the air in the chimney, and more air is drawn into the chimney from the inside of the structure below. It is then heated, and the cycle continues.

Any fireplace or wood-stove chimney will convect a certain amount of air from a structure, but it is probably the least effective method. Clerestory windows (which are not usually considered to be chimneys), do a remarkable job of extracting unwanted air from a structure, and in many cases are advisable for this purpose alone. Skylights and sun scoops are also good places from which to exhaust unwanted warm air for ventilation purposes, but care must be taken to insure that waterproof conditions are not disturbed.

Knowing what I do now about well-planned floors, I would just level the floor soil as best possible, put down insulation board, and lay brick directly on the boards. If moisture is a problem, I would use a sheet of polyethylene under the whole thing. In some areas I would just use the sheet of polyethylene and then carpet it over, not using any insulation or brick at all. For some floors such as those in bedrooms, baths and the like, I would put in a hardwood floor. Notice that all of these floor suggestions involve material that is soft and tends to give when walked upon; however, hardwood floors just cannot be beat for comfort and appearance, though they are expensive. By using a soft system, all pipes, wires and utilities are more accessible.

In Mexico and the southwestern United States, adobe homes use the earth as a floor. Some of the floors are treated by sweeping linseed oil into the soil repeatedly until it becomes almost as hard as concrete, while still maintaining some give. A more artistic approach is to mix adobe and pour it six inches thick on the floor, wait weeks for it to dry, then crack it into a mosaic. Another mix of adobe is then swept into the cracks. When this dries, linseed oil is swept on and allowed to dry, followed by another coat. The floor hardens like concrete, looks like tile, but still has plenty of give.

Advantages of a one-level home

I designed the Prairie Cocoon as a one-level living space since I feel that the most useable space is achieved this way. Resale is very important when trying to get a loan to build, and a single-level home seems to have more appeal to all age and income categories. The elderly can move through the structure with a walker or wheel chair. It is not only the elderly who would appreciate this convenience: If you have ever broken or sprained a leg, climbing stairs is a real trial.

Plants can be strategically placed to create a multilevel effect.

One level is convenient, but it can look boring, both visually and physically. This is one area to which I would pay more attention were I rebuilding. I would still retain the single-level concept, but I would have changes of visual levels. The visual monotony is more noticeable than the need for physical movement to different levels. Achieving visual change is fairly easy. One of the best ways to do this in the case of the Prairie Cocoon would be through the use of plants and platforms. The living room could have been divided into specific areas of privacy. Low hedges in planting boxes or bushy plants would have provided another level. Trees would have added another, and platforms with plant arrangements would have broken the horizontal appearance of this room. More about visual levels will be covered in detail under the chapter on landscaping.

Disadvantages of the Prairie Cocoon

There were two fairly imposing problems that I encountered in the Prairie Cocoon that are common in many underground structures: high humidity in the summer and the accumulation of mould and mildew in some of the closets and back rooms. Were I to rebuild, the skylights in the back would take care of some of the mould and mildew problems since neither of these fungi can grow when exposed to direct sunlight. Good air movement and lowered humidity would

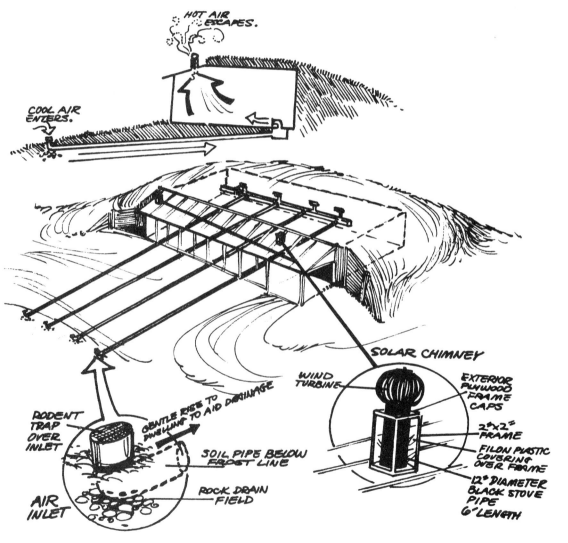

This soil pipe and thermal chimney system would have prevented problems with humidity, home cooling and fungal growth.

have alleviated the problem, since humidity and cool temperatures encourage the growth of these fungi. To accomplish this on a rebuild, I would incorporate solar chimneys on the greenhouse and install a system of soil pipes. The chimneys would draw the heat and stale air out of the structure, and the soil pipes would supply cool, dehumidified air into the structure. A more complete breakdown of how this system operates is covered in Chapter 7.

Since closets are always dark, a small ultraviolet light in closets where fungi persist would stop them from growing. Once the pungent odor of fungi is established, it is very hard to eliminate. We used every kind of cleaning solution on the market and were still unable to defeat it. If the systems just described had been in place, there would not have been a buildup in the first place. The larger the buildings are, the more prone they are to this problem since there will most likely be more dark or unused areas. The main ingredients in avoiding this situation are to keep excess humidity out of the air and to be sure floor and wall areas remain dry through proper waterproofing and vapor barriers.

The Prairie Cocoon could have been built using different materials and techniques. The labor, materials, transportation and residual costs were all weighed carefully before arriving at this particular formula. Such methods and materials as tip-up wall sections that are cast on the ground and a poured-in-place roof could have been used. This same structure could have been built entirely out of wood, using the post-and-beam system. These options and others are usable on many underground designs.

7
Trench House

A replica of an early dugout, which would be your home, need not be an uncomfortable ugly structure. Indoor plumbing would not be authentic, but old fixtures such as a clawfoot bathtub, pedestal lavatory and an old kitchen sink would still give the home a nice appearance without removing its period flavor. Additional glass for light could also be added without destroying the dugout concept. A lot of warmth and charm was already inherent in the early designs, and all you need to do is to add some convenience and comfort without changing the basic design. The following design—the Trench House design—is similar to many of the early dugout designs that used basic post-and-beam construction. This may seem too primitive to put in a book on modern underground construction, but it may be just the thing for the person or couple desiring basic low-cost and energy-efficient shelter. This design would be especially suited to the park concept of acquiring land or for the homesteader who wants to get back to the land and be as self-sufficient as possible. It is also an excellent design for a young couple or single person with limited funds.

Legal aspects of building a replica of a pioneer dwelling for occupancy will vary according to local restrictions. A small incorporated town would have control over a four-square mile area, which might fall into a special use area and be exempt from the majority of restrictions.

This pole-supported, shed-roof dugout can be very comfortable and light. The only concessions to our modern methods and materials is going to be electricity for light and refrigeration. Other conveniences such as plumbing and fixtures can be carefully selected so as not to detract from the feeling and appearance of the structure's historical concept.

The design to be presented here presumes that you will be able to cut into a hill; however, it does not have to rely on a south-facing slope. It is a rectangle that requires a trench cut that is similar to a farmer's silage pit. The dwelling's 14 × 40 feet are within the limited scope of most early dugouts. The floor plan provides a feeling of openness while still providing privacy areas. Some of the earliest dugouts were comprised of only one room, and a dugout that was 12 × 12 ft. was not that uncommon; one account in the Nebraska Historical Society's files tells of a family of six living in a dugout this size.

Excavation

Construction begins with excavation. If you are building the structure as a historical demonstration, I recommend that you do your digging by hand. Volunteer labor for this might be available through the channels mentioned earlier. A trench cut can be dug from both ends and, consequently, two digging crews can

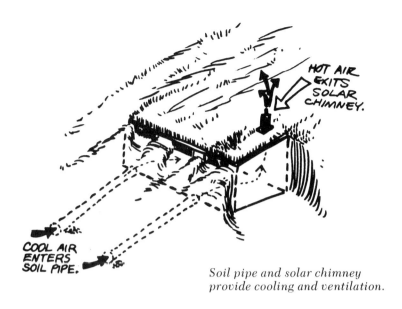

Soil pipe and solar chimney provide cooling and ventilation.

cut the digging time in half. Spades, shovels and wheelbarrows will work just as well as power equipment for this type of cut. Dig the trench about two feet wider than the planned structure; a sixteen-foot-wide trench will work fine.

A decision has to be made at this point as to whether the structure will be built as a log structure or with posts and lumber-shoring. Pioneer dugouts in Nebraska, Kansas and North and South Dakota used both systems. The determining factor was usually whether or not a sawmill had been established nearby. Most of the later dugouts at the turn of the century were post and shoring. Logs cut along the creek banks were used initially because they were available, and the creek usually supplied the settler's water. Slopes found along the creek made the job of building a dugout with reasonable drainage easier. The pioneers did not construct the courtyard type of dugout on the Great Plains of the United States because drainage was a problem in the clay soils, and the winters would fill the excavation with snow. The main reason the early dugouts tended to be dark was due to the fact that glass, when available, was expensive. The Trench House design will make some concessions to the glass problem and include several skylights for even light distribution. The glass will be in panes, though, to keep as close to the pioneer design as possible.

Setting the posts, girders

Once the evacuation is completed, posts will have to be set as retainers for either the log or lumber sheeting. The ends of the excavation should be angled back downhill to allow drainage and to help protect the ends of the dwelling from wind. Post retainers will be needed for these cuts as well. The log sheeting will not need as many retainer posts as the board sheeting. Eight- to twelve-inch diameter logs will wall up the sides nicely. Four posts, ten feet apart, will hold the logs in position for back filling; however, shorter spans may be necessary for roof-support girder placement. The logs should be stacked without staggering so that the retainer posts can be built with limited openings, and then near the top, thus simplifying construction and shortening building time.

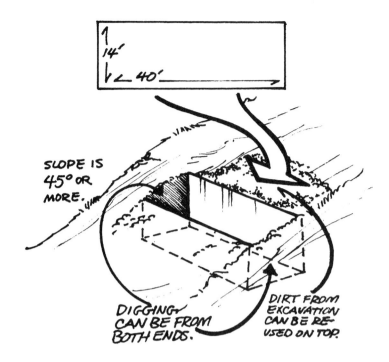

Excavation for the Trench House can be accomplished by hand or machine with relative ease due to its open-ended design.

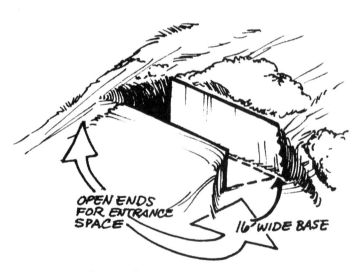

Set the corner posts first and then run a string around the shoring side of the posts on the outside of the trench. The other posts will have to be dug in and lined up on the string. Since all of the posts will vary slightly in diameter, this assures that the wall will be straight on the outside. The setting of the posts should be done with great care since they not only have to support the walls against the expansive lateral and vertical pressure, but they also have to support the compression load of the earth-covered roof.

Since these posts will be in the ground and subject to constant moisture and decay, they need to be protected. I am against treating posts with insecticides and preservatives, since I feel these chemicals always have residual effects on all

	Length L (feet)	Rafter Round (dia.)	Rafter Rect. (nom*)	Girder Round (dia.)	Girder Rect. (nom*)	Post H=approx. 10' (diameter)
a)	4'	5"	3×6 (2×8)	8"	4×10	5"
b)	5'	6"	4×6 (2×10)	8½"	6×8 (3×12)	6"
c)	6'	7"	4×8	9"	6×10 (4×12)	6"
d)	7'	7"	6×8 (3×10)	9½"	6×10 (4×12)	7"
e)	8'	7½"		10"		7"

* MEANS NOMINAL DIMENSIONS SOLD BY LUMBER STORES. TRUE DIMENSIONS ARE SLIGHTLY SMALLER. TABLE ALLOWS FOR THIS VARIABLE.

Figures above are load figures based on an eight-foot-thick soil cover and apply to other post-and-beam structures in addition to the Trench House.

living things and are apt to get into the air and may be subject to human contact. Replacing these posts every year would be a real headache, so protection of some kind is necessary. I would coat the part of the posts that go underground and a few inches above ground with liquid butyl or one of the new plastics that stretch without separating. These protectors can be painted on when in liquid form and then set up in an elastic surface that never hardens but provides a moisture and vapor barrier.

The main reason that some codes do not encourage construction of a pole building is because the posts that support the structure tend to rot. Many chemicals that are used to pressure-treat lumber used in wood basements that meet code requirements contain dioxin—one of the most deadly poisons ever created. As an additional barrier against decomposition, the posts could be wrapped in a couple of nonbiodegradable garbage bags.

Setting the posts in their holes should be done with care so as not to tear their moisture barriers. A fieldstone larger in diameter than the post could be placed at the bottom, with the post set on top. This would serve as a base and help spread the load. The rock would keep moisture from collecting at the bottom of the post. A couple of shovels of crushed rock or large gravel at the bottom of the hole will help set the post while helping to drain water away from it. The posts should be cut extra long so they can be topped to the correct length for setting girders.

Post retainers in the trench make construction easier.

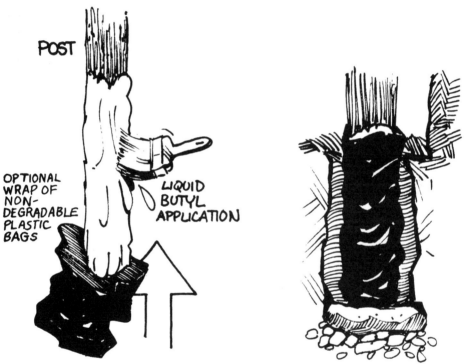

Nontoxic protection is applied to support posts; careful setting of these posts keeps the waterproofing intact.

Placing shoring logs

Once the posts are set, the next step will be to place the shoring against their outside. Logs will be used in this first example, then we will discuss the use of mill ends later in the chapter.

A trench should be dug on the outside of the support posts, in line with the vertically dug walls. The trench will in effect lower the outside of the excavation below the floor level of the inside of the structure. This trench should be filled

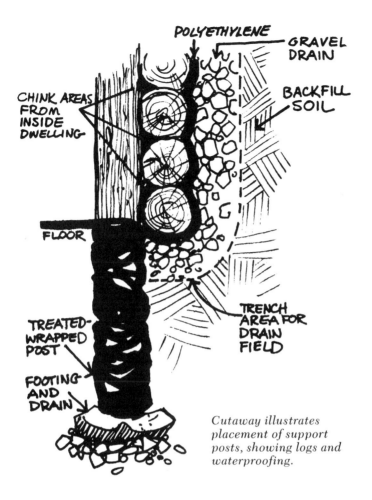

Cutaway illustrates placement of support posts, showing logs and waterproofing.

with a couple of inches of gravel so that the top of the gravel is at least two inches lower than the inside floor. Polyethylene should then be placed under the first log that is placed on the gravel and against the support posts. The plastic should extend at least a foot up the inside of the wall, lapping onto the floor. This will protect the shoring log's underside and keep moisture from its outside that will be backfilled. Two or three sheets of polyethylene would be better than one. Great care should be taken to assure that the polyethylene is not ruptured during the process of lifting the shoring logs into place. The plastic is the only barrier against moisture and vapor, so it has to be kept intact.

The polyethylene can be rolled up the outside of the logs as they are stacked. Backfill should be accomplished as the logs are stacked, allowing gradual compaction and settlement. At least four inches of a coarse gravel should be placed against the polyethylene, and soil should be backfilled against the gravel. This will create a gravel drain field against the outside walls and eliminate the possibility of standing water. The base gravel under the bottom shoring log should extend beyond the ends of the structure so that infiltrating water will escape the drain field and get away from the structure entirely.

There will be some fairly large openings that will show between the logs. These openings should be filled on the inside with chinking. If the outside surface of the log shoring is too uneven and the polyethylene may separate during backfill and settling, a chinking of mud on the outside will even out the surfaces. Let the mud dry completely before covering it with the polyethylene. If the mud is covered while wet, it will dry only by absorbing into the wood, causing the wood to rot.

Before too many shoring logs are placed, the top girders will need to be notched and bolted into place. These girders not only support the entire roof system, but they also stay their opposite member supporting posts. One of the real advantages of the Trench House design is the backstay each side of the trench exerts on its posts, which in turn receive the earth's thrust from the opposite side. Diameters and spans of the various members of the post, girder and beam-and-rafter system must be considered under the actual conditions they will be expected to endure. Remember that the loads imposed will be permanent. There is some safety built into the figures, but the choice of logs will have some bearing on the strength. Certain woods bend while others break; some woods take compression better than others; and some woods have a hardness that makes them less likely to give. The engineering chart and the chart on various strengths and qualities of woods will help guide you in the wood selection process. Each piece should be looked at with an eye on how it will be used in the structure and what forces the wood will have to withstand.

Even though many building codes do not encourage the use of pole structures, such structures were originally developed by the Japanese to withstand earthquakes. The poles were able to move individually, then settled back into position with little damage to the structure, while a building on a solid foundation broke up with the foundation during an earthquake.

Lighting

Light is admitted to the middle of the structure through pie-shaped openings that gather light into the exposed windows of the downhill wall. These openings are created between supporting posts and require little additional work or expense, and they also benefit the interior with a great amount of light. These windows and the end wall of the elongated dwelling should be comprised of small panes to keep the feeling of antiquity. These windows can usually be bought at salvage yards for very little cost, and the individual panes are very easy and inexpensive to replace in case of breakage. These high roofline windows barely break the surface of the downhill slope continuity, and if properly installed, will not be visible unless someone is right near the window itself.

The primary idea behind the Trench House design is to offer the least amount of exposed structural surface as possible to the hot winds of summer or the cold winds of winter. The dwelling's profile offers the best protection possible in case of tornado. The design also lends itself to natural drainage in a downpour and the least amount of heavy snow load, thus limiting additional load on the structure. Conduit for internal wiring should be fastened to the outside of the walls before backfilling takes place.

Constructing the roof

Once the shoring is in place and the drain field and backfill are complete, the roof system is next on the agenda. The system is made up of girders, rafters and planking. The diagram explains each component's function. Basically, the girders support the roof while staying the posts that support them. The rafters support the planking that in turn supports the earth load of the roof. Each part of the system

Placement of conduit on the exterior of the house prior to backfill.

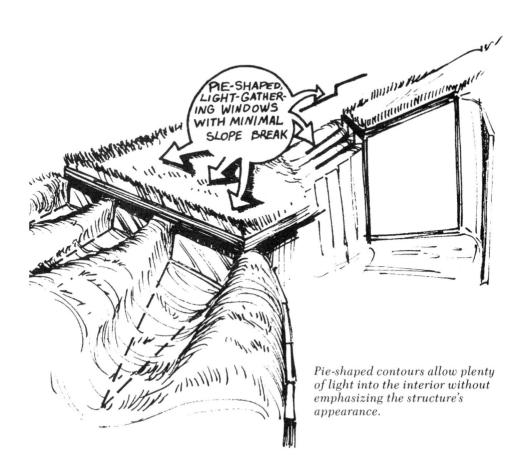

Pie-shaped contours allow plenty of light into the interior without emphasizing the structure's appearance.

The Trench House offers no wind profile along its downhill drainage.

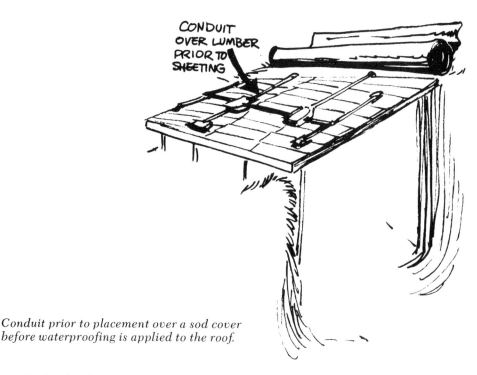

Conduit prior to placement over a sod cover before waterproofing is applied to the roof.

spreads the load to another stronger member, which eventually becomes the great compression strength of the posts. Even a 2 by 4 will support about ten thousand pounds of compression load, but its slenderness ratio would make it vulnerable to shear if struck from the side. The thickness of the supporting posts has more to do with the possibility of side shearing than with compressive strength.

The girders, rafters and planking have to be able to bend without breaking easily and, consequently, are chosen for this ability to bend. Since the girders should already be in place, the next step is to place the rafters, which will have to be spaced accordingly to the type of planking used. Poles or small-diameter logs may be used for planking, since we are presuming the more primitive pioneer log dwelling in this description. These logs may be four or five inches in diameter, and in order to give the roof a reasonably even surface for drainage of the gravel field, you must fill in any spaces between the logs. This filling can consist of clay that is allowed to dry before the polyethylene roof sheeting is applied. Conduit for any ceiling fixtures will have to be run prior to chinking. The conduit can be placed in the valleys between the planking logs or poles and then chinked over.

Flooring

The cost of this structure can be kept at a minimum if the tile-like dirt floor described earlier is used in this dwelling. It lends authenticity to the design, while also being attractive. The cost of floors can really get out of hand, and using a tile-like dirt floor is an excellent way to cut a major cost without giving up comfort or appearance. A dirt floor is very easy on the feet and accessible when installing plumbing, sewers and utilities.

Using mill ends

Building this structure using mill ends or lumber shoring is no different than using log shoring to construct this design, except the space between support posts and rafters will have to be shortened. Lumber shoring will have less shear strength than that of logs or poles and, therefore, will have to have extra support. Recycled 2 × 12s in good condition from a salvage yard would make excellent sheeting. When using this heavy lumber, the rafters may be spaced about three-and-a-half feet apart. If three-quarter-inch lumber is used, the rafter spacing will have to be kept at about two feet. Longer lengths of lumber used for shoring are probably stronger than short pieces, but short pieces may be more functional. Your rafters will not be exactly the same diameter, so short pieces will nail on centers more easily.

The use of logs for girders and rafters means that the roof will tend to be somewhat uneven. As long as ridges are not created on the downhill slope of the shed roof, water will drain from the roof. Here again, the conduit has to be placed for ceiling fixtures prior to putting the polyethylene sheeting and fill in place. The lumber sheeting will not leave valleys in which to place the conduit, so be sure to make the conduit run in the same direction that the runoff will occur; otherwise, your roof will have a series of mini-dams that will collect water which will eventually find its way inside your home.

Framing

Before carpentering in the ends of the structure, make sure all of the pipes, wires and chases are in the floor and run to their outside destination. The end walls will require very little framing other than a few 2 × 4s to serve as stops for the multi-paned windows. One glass layer on these ends will not be an absolute barrier against the cold, but because of the way the glass protects against windchill, you will be dealing mainly with surrounding air temperatures that do not penetrate the same as wind-driven cold. The extra mass of earth around this trench design also will help offset these two noninsulated walls.

Greenhouse temperatures

The entire greenhouse on my first house was extruded. All of the experts said that I would lose too much heat at night if I did not put up movable insulation. I had felt that the tempering effect of the soil surrounding the main structure would offset any heat loss through the nearly four hundred square feet of glass, and it turned out that I was correct. Even during construction (before the greenhouse ceiling had been insulated or the doors installed), the coldest it ever got was 45°F, and that was with outside temperatures of five degrees below zero. The mass of the two long walls and the wind-protected ends should net an inside winter temperature of nearly 65°F once the structure has stabilized. The small temperature differential needed for comfort on extreme occasions can be easily met with a wood-burning stove.

Solar chimneys and ventilation

Soil-pipe ventilation and dehumidification in conjunction with a solar chimney will make this dwelling more comfortable than will air conditioning. The diagram shows how the above system would work in this particular building. I recommend the use of soil-pipe and solar chimneys on all underground structures (as well as on surface structures). When using this system, be sure that the hot air has a hole from which to escape, and the cool air drawn through the soil pipe (by the siphoning effect of the escaping hot air), has a hole through which it can enter the structure. The two holes should be at opposite ends in the structure to allow the cool air to be drawn through the structure on its route following the escaping hot air.

Miscellaneous tips

Some additional tips regarding the construction of this dwelling will complete the section on this particular design.

1. The windows on the ends of the structure do not have to be the same size. The random sizes can be mixed very artistically and actually lend to the authentic pioneer appearance.
2. Use leverage whenever possible while lifting the various heavy framing members into position. I found that a simple post winch that farmers use to stretch fences will lift any heavy piece into place using a boom that is made from one of the logs.
3. Take your time with backfill. Do the backfill by hand, and allow it to settle after tamping. Pouring water on the fill as it is placed in two-foot lifts and allowing each lift to dry will assure very little settlement upon completion of the structure. This gradual process allows all of the supporting frame members to receive an adjusted load. Fill both sides of the long walls at the same rate so that no uneven loads are imposed on the frame to cause it to torque or twist to one side.
4. The earth floor should be constructed as mentioned previously, and the only two floor areas that require a hard concrete base are the bathroom and garage, and since this design does not include a garage, you need only to hard-surface a very small area: the bath.
5. All of the pieces of this design do not have to match up exactly; the pioneer dwellings did not either. Mud, plaster or an extra board will usually suffice, the result often being a visually attractive enhancement.
6. To break the monotony of this rather tunnellized structure, you can add a level change in the living area. This can be done by simply digging a conversation pit. The pioneers sometimes did this to provide sitting places on the edges of the pit, since furniture stores were few and far between. (Beds were sometimes created the same way by simply carving out a raised bed.)
7. The retainment walls at either end of the house can be painted white to

increase light reflection into the structure when the sun does not radiate directly on the glass, as can the small retainment walls that form the pie-shaped window openings on the top of the downhill wall. The roof overhang will cut direct heat gain on this glass in the summer, but it will also darken the area. The white window wells will compensate for the shade by reflecting the light without appreciable heat gain.
8. A root cellar carved in either bank would be an interesting addition to this dugout design and serve a practical purpose as well. Many of the foods that we refrigerate could be kept in the dry coolness of a properly constructed root cellar. By tunnelling directly into one wall near the kitchen, you can have all of your garden produce within a few steps. For more information on building a root cellar, see Chapter 23.

8
Railroad Tie House

The Railroad Tie House is an economical, basic, nearly self-sufficient form of housing. This design, along with some others in following chapters can make excellent weekend or summer homes or hideaways, and become economically feasible when the cost of vacation and travel is taken into consideration.

The Railroad Tie House is comprised of 560 square feet of floor space, which may seem like a very small living space, but once you accommodate your thinking to this space, it is more than adequate for two or three people. It is another post-and-girder-system design, but it uses a commodity that is fairly common in most areas—railroad ties. I chose to use railroad ties in this design because there are areas where trees and lumber are not readily available at prices the bargain-builder can afford. The Great Plains area of the United States is not overloaded with pine, cedar and fir trees, so I turned to railroad ties as an alternative. The supply of ties may not last too long, though, since some railroads are using cutting machines that chop the ties in pieces prior to removal from roadbeds. Most of the areas where I travel still have large supplies of good used ties at reasonable prices. You might perhaps work out an agreement with yard foremen or section foremen for the salvage of these items.

This design is equally attractive whether used as a hill home or as a totally bermed structure. Modular in concept, it can be made even smaller than the plan calls for—or larger than the plan, if you so desire. The idea of building it initially as a weekend cabin and then adding on in the future may be especially appealing to those who have had no experience with underground building or living. The cabin version could be built without electricity or extensive plumbing. Utilities could all be added later if accommodation for them was made during construction. This simpler method would cut costs even further, while giving everyone concerned some positive experience with earth-shelter building and living.

Plumbing

A well that's driven on the building site would cut initial plumbing costs, and a small hand pump mounted on a sink directly over the well would be helpful. Such a system would provide drinking, cooking and bathing water. A composting toilet would eliminate the need for sewage lines and the additional septic system or lagoon. Bath or shower water could be heated while cooking on a wood- or cob-burning stove, or it could be heated by an economically built thermal-syphon solar system.

Construction

The entire cabin is designed in eight-foot sections, using standard railroad ties—with the exception of the vertical posts. Vertical posts will have to be fourteen-foot yard ties, though twelve-foot yard ties could also be used. A total of nineteen yard ties and 252 regular ties are needed in order to complete construction. Included in these figures are all dividing walls and the roof. Railroad ties used in this design are for girders; therefore, rafters are not required since each roof plank is a tie that is self-supporting. Since the railroad ties make up 95 percent of the building material for the structural shell, this is a very inexpensive method of building, especially if you can get a discount price for buying by volume.

The Railroad Tie House is put together much the same as the Trench design. Set corner posts and fasten a string between them on their exterior side. Set the other posts to this line to assure a straight wall. Notch the overhead girders and place them end to end over the posts. The posts should be set in the same manner as described for the Trench House. Even though the creosote should protect the post ends, the use of garbage bags around the post ends will ensure it.

Some people have expressed concern regarding contact with creosoted ties and their odor. I have checked with the treatment facilities of a railroad and was assured that there was no insecticide mixed with the creosote. Creosote taken internally in any quantity would be fatal, but there are no toxic air-borne fumes. The use of old ties also eliminates most odor possibilities since ultraviolet rays have eroded most of the surface and near-surface treatment. A number of paint-on plastic sealants are available that would also keep the surfaces out of physical contact.

For added strength and eye appeal, I would trowel on a surface coating of a surface-bonding material comprised of cement and fibreglass that is used for bonding dry-stacked cement blocks. Metal lath or chicken wire could be nailed to the railroad ties, and the bonding agent would then better adhere. This would leave a textured white surface on the inside which would be both attractive and light-reflective. Note that creosoted railroad ties can be safely used, but their surfaces must be thoroughly protected from human or animal contact.

Installing skylights/windows

A south-facing glass wall is needed for heat gain in winter. Skylights are easy to install since all that is involved is removing two ties, building a box in the opening and installing glass on the top of the box. I would hinge the glass covers so they can be used as solar chimneys to ventilate the dwelling in the summer. These skylights in the back half of the fourteen-foot-deep interior will balance the interior light. This is very important since the light on the glass wall in the front will be very intense, especially in the winter when the sun is low. The glass for the wall can be composed of multiple-paned windows that can be found at a salvage yard, since double-paned thermal windows are very expensive.

It does not really matter that all of the windows do not match in size. An artful arrangement of the windows will give a very pleasing visual effect. The only problem with this design is that there is not enough length to the roofties to make an overhang that will sufficiently shade the glass during the intense heat of the summer. To compensate for the missing overhang, a snow-fence roof trellis is attached which extends from the front roofline to posts that are set ten feet out from the front of the house. Morning glories or other vining plants can be planted at the front edge of the roof and trained to grow over the trellis. Not only will the trellis help shade the glass, but the plants will add their cooling transpiration to the area below.

The Basic Earth-Sheltered House

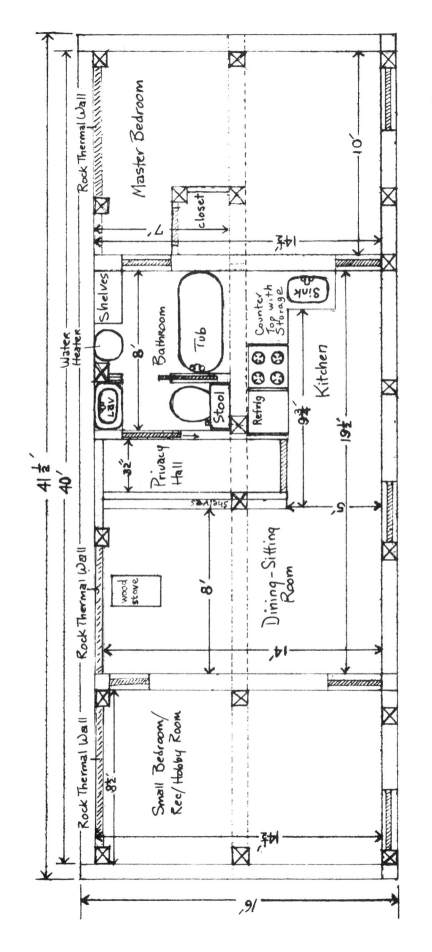

Opposite: Railroad Tie House. Above: Floor plan for the structure illustrates a layout for a 580-square-foot house, with an interior of 14½ feet by 40 feet. This design is modular and can be added to in lengths of eight feet. Be sure to seal ties by covering them with a vapor barrier and other inner-wall material to avoid any contact with creosote.

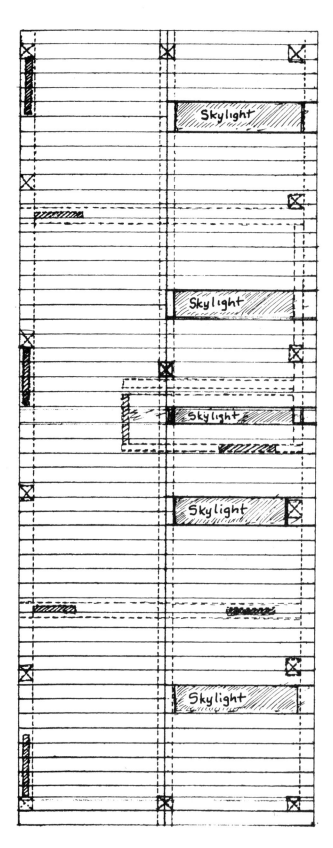

The roof section of the Railroad Tie House shows the placement of railroad ties in relation to support walls, center beam system and wall support posts. The roof system is constructed entirely of standard eight-foot railroad ties and is stressed for a one-foot earth load. The roof should have a ten-foot drop front to back, with roof and side walls surrounded by a four-foot gravel field for drainage. Skylights are box-type and are open for ventilation; they serve as solar chimneys when combined with the soil pipe cooling system.

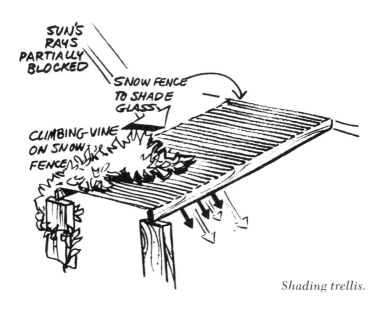

Shading trellis.

Flooring

Not only have you helped to shade the windows, but you also have created an ideal sitting patio, which can have random brick (also salvaged) as its surface. Interior floors can be covered similarly by placing polyethylene on the floor and then laying bricks over this vapor-and-moisture barrier. Only the bathroom area would have to be grouted. The natural give of the bricks make it a floor that is easy on the back and legs. Cleaning up a spill is no problem at all, since you can just pick up the bricks around the spill, sop up the spill, and then replace the bricks. The natural grime that is a part of every house will work its way into the cracks and grout them for you.

Building sequence for the Railroad Tie House

1. Excavate a hill cut that faces south, or start with flat ground and simply level it. The dwelling on flat ground will be bermed and covered later.
2. Set corner posts, snap a line on the outside, and set the other vertical posts. Be sure that the posts are set the correct distance apart so that the ties that are being used as shoring will butt in the center of the outside of the posts.

Post setting.

3. Notch the girders and fasten them to the tops of the posts. The ends of each girder tie should be notched about two inches deep and six inches back. This will allow the girders to butt together on outside walls at the top of the support posts. Unlike the trench design, this design uses only girders and posts; the sheeting on the roof is comprised of ties that are also rafters. This system requires that the girders run around the perimeter of the structure and down the center so that the sheeting rafters radiate from the front and back girders into the center of the structure and rest on the center girder system. Everything is thus divided into neat eight-foot squares.
4. After setting the girders, fasten them to the tops of the posts with strap-iron brackets. Bracketing keeps them from "jumping" up out of their notches, causing structural failure if one of the posts should shift due to an uneven amount of thrust from the backfill. This bracketing and notching will assure a strong structure.

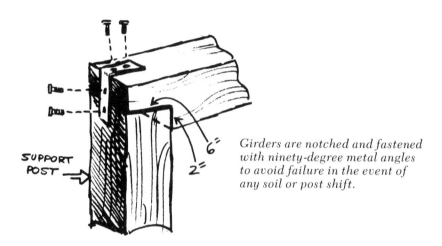

Girders are notched and fastened with ninety-degree metal angles to avoid failure in the event of any soil or post shift.

5. Start placing the bottom shoring-ties around the three outside walls. Go up about three ties on the three sides. Be sure that the trench is dug for the bottom ties and the drain field. This trench will be under and in back of the ties. Also, be sure that the polyethylene has been wrapped around the bottom ties and laps the inside floor about a foot. If the trenching has been done correctly, the inside floor should be about three or four inches higher than the bottom of the outside trench and one or two inches above the drain field gravel. The gravel field should exit at the front around both ends of the structure to assure that all of the saturation gets away from the building.
6. Continue the shoring up to the roof-and-girder line. Be sure that the two or three layers of polyethylene are brought up and backfilled with four inches of gravel, followed by a sop berm large enough to cover the structure. Do this fill in two- or three-foot lifts. Two-foot lifts are better and assure proper settlement if the fill is irrigated and allowed to dry as each lift is placed. This takes longer, but will guarantee even backfill thrust on all of the shoring and support posts. It will also assure that sudden settling does not tear the polyethylene, which is the only raincoat your house has.

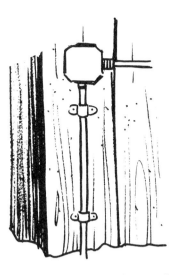

Conduit and wiring should be secured to the exterior prior to waterproofing and backfill.

NOTE: All conduit for internal wiring outlets should have been fastened to the roof and outside walls prior to waterproofing and backfill operations. If it was not accomplished at that point, all the wiring will have to be fastened to the inside walls and, consequently, will look very unattractive.

7. Next, lay the roof-sheeting ties. If large cracks appear between these ties, mud them in order to level the surface. The front girders should be well anchored since their posts have no backfill to give them support against thrust from the opposite side.

Filling roof cracks.

8. The two main dividing walls can be constructed out of railroad ties or from used 2 by 4s and any sheeting you feel appropriate. The beauty of this design is that the interior walls are nonsupporting and, therefore, can be moved or eliminated, if you choose.
9. The front girders should be about four inches higher than the rear wall girders to give the roof a slight pitch to the rear for drainage.
10. Once the roof surface is as even as possible, put down the polyethylene sheeting. Two or three layers are preferable. Be sure that these sheets overlap the sidewall sheets by at least a foot. Tape them with plastic tape to assure that water does not capillary back up between the two layers.
11. Put the roof gravel field on; backfill. Do all of this by hand and be very careful not to snag the plastic sheeting while levelling or moving the dirt and gravel around. If a rake is used, use the back side of it instead of the tines.

12. This system will carry two feet of earth, but it is my feeling that one foot is more than enough to grow grass and serves the purposes of tempering, transpiration and roof protection.
13. The shed roof can have its pitch from front to back or back to front. If the house is built into a south-facing slope so that the cut at the rear wall is the same height as the wall, I would slope the roof towards the front to follow the contour of the hill. Be sure that the front patio has a slight downhill slope away from the structure, too, in order to get rid of the roof runoff as quickly as possible. One thing about a good sod roof is that the runoff is not nearly as great or as rapid as from a conventional roof.
14. The framing of the front window wall can be out of recycled 2 by 4s, 2 by 6s or whatever is available at reasonable prices. This wood can be stained or painted to improve its appearance. I saw one front of a structure done in old lumber that looked attractive in its natural weathered state. Needless to say, the framing will have to follow the window-frame sizes that you acquire.
15. Set the patio posts. These can be railroad ties or just about anything that you like since they will not be bearing any real load. Two by sixes can be run from rafter hangers nailed to the front girder system to the posts in front of the patio. To keep the 2 by 6s from being knocked off, they should be bracketed to the posts.
16. The snow-fence trellis should not be permanently attached since you will want to take it down for better heat gain in the winter. I would advise attaching it with screw hooks at both the house end and the post ends. Morning glories are a very fast-growing vine and will cover the trellis in a matter of weeks. This system is super-attractive, inexpensive and functional.
17. Retainment walls will have to be erected to stop erosion next to the front of the structure and to form drainage wings away from it. Railroad ties work well for this purpose. If the structure has been built on flat ground and the fill has to be brought in to berm it, the berming can be handled one of two ways. Terracing is one way, making a large gently sloping mound is the other.

    It is hard to get grass started and a root system established on a gently sloping mound before water erosion does a lot of damage. I would build the berm out of terraces that are retained by railroad ties. These ties can be floated singly without anchoring, and the terraces will be shallow. The shallow terraces will eventually be unnoticeable due to settling and grass growth. Allow the grass to grow as long as possible, since it serves as insulation both in winter and summer. I prefer to use grass that is native to the area. My first house was mostly bermed, and it blended so well with the surrounding area that no one ever noticed it until they drove around in front of it.
18. Some last construction tips should be included here as a conclusion to the building steps of this design. Be sure that not only is the conduit for ceiling light fixtures in place before waterproofing and backfill, but be sure that flues, vent stacks and any other openings are included. To keep the skylights from leaking, use galvanized sheet metal over the top edge of the box, down the outside and over the polyethylene roof sheeting.*

---

*The box is made to line the opening in the roof to support the skylight.

Apply plenty of liquid butyl or other nonsetting sealer around the base and anywhere else that it might capillary.
19. Run the soil-pipe system into this structure by bringing the pipes uphill into the front. I would use four six- to twelve-inch diameter pipes. These should vent into the front of the structure's floor. A ceiling vent such as a cracked skylight will work just fine as a solar chimney. Experiment with opening the pipes singly or several at a time. An opening that is too large will not draw as well as a smaller one. You can regulate air flow to suit your own comfort this way.
20. The interior of this structure can be handled in a variety of ways, depending primarily on your tastes. For wall and ceiling surfaces, I prefer using bonding plaster that is then painted white. This light-reflective surface is nice, and just about any furnishings or colors can be used in contrast to the white. I like either handmade furniture, similar to the Shaker designs, or antique furniture. Wooden furniture, combined with the plastered interior and brick floor, would give this design a feeling of antiquity while also giving it great warmth. Lush, green hanging plants would be good finishing touches.

9
Deep Woods A-Frame

The Deep Woods "A"-Frame was designed for mountainous areas that have an abundance of timber. As you can see by the height of this structure at its peak, opposite, it would require a lot of backfill if it were built on flat land. A steep slope that you can cut into, preferably near the top of the slope to minimize slope shift and earth thrust, is the type of terrain ideally suited for this design. The design is essentially a buried A-frame home that can be scaled to fit the size desired by the builder. The Deep Woods A-Frame is also suited to the needs of the self-builder with limited equipment. Its strength comes from heavy logs that are fastened to a ridge plate, with the second-floor deck and support members adding reinforcement at the center. The skylight and glass on the south-facing front wall provide needed heat for winter. Skylights help to provide balanced light to both the first and second levels.

Building techniques required for this structure are simple and straightforward. The logs that make the "A" can be lifted into position with a tripod (a lash-up of three poles, 15 feet in length and five inches in diameter) and block and tackle (which lifts the logs in place and is suspended from the apex of these three poles) after placing their butts in a trench for backstay. This design can be put together for very little cash, since the shell of this structure is nearly free. A government cutting permit, a chain saw and a way of hauling the logs are going to be the main expenses. The back end of the "A" (which is buried), is made of logs placed horizontally as shoring on a post-and-girder structure. The ends of the shoring logs will extend to the outside of the roof logs so that they are stayed by the entire structure. Two additional posts at even distances across their span will also keep them from being pushed inward. Additional engineering information will be given in the building sequence.

Generally speaking, the A-Frame is not an energy-efficient design. When built from conventional lumber, a great amount of insulation is necessary. Also, the heat in winter tends to pool at the peak, leaving the bottom floor cool. When using the A-Frame as an underground design, the utilization of heavy log construction eliminates these undesirable aspects of the design and gives it a unique charm.

The following sequence of construction steps will provide the basic data needed for you to fully consider this design's application to your own needs. None of my figures and suggestions should be taken as absolutes because every building situation has its own special set of problems and conditions. The use of a qualified engineer as a consultant on any design, whether mine or someone else's, is always advised. Qualified engineering help can be obtained at reasonable rates from a variety of sources. One of those sources may be the engineering college nearest to your location. Many companies that sell structural building

Elevation of the Deep Woods A-Frame.

components retain engineers as part of their staff to be sure that the components they sell are used properly and within their design limits. Many of these engineers freelance, and their fees are very reasonable.

Construction steps

1. Excavate according to the size of the planned structure. Since this design will most likely be built in mountainous terrain, there may be a need to do some blasting in order to remove dirt efficiently. Logging companies usually have a number of employees who are experienced at blasting, and you might consider having a skilled person do whatever blasting is necessary for your excavation.
2. Dig a trench around the perimeter of the excavation. The back trench and drain field serves the same purpose on this design as it did for the Trench House and Railroad Tie House designs. The two long sides of the excavation not only serve as a drain field, but also as a "kicker plate" for the long logs that make up the sides of the "A." The side trenches should be lined with fairly large rocks to help spread the downward and outward thrust of the A-Frame shoring logs. These large rocks will help drain the log ends and reduce the possibility that the shoring logs may rot.
3. Cut the logs that will serve as the shoring for the sides of the A-frame. A bottom base that is twenty feet wide will require logs that are about twenty-two feet long in order to allow a ten-foot ceiling height for the lower level and a ten-foot peak height for the second level. As long as these logs are about the diameter of the average utility pole, they will support the loads that will be imposed on them.
4. Set the posts for the post-and-girder support system for the second level. These posts will have to be set and their girders positioned to support the centers of the shoring logs.

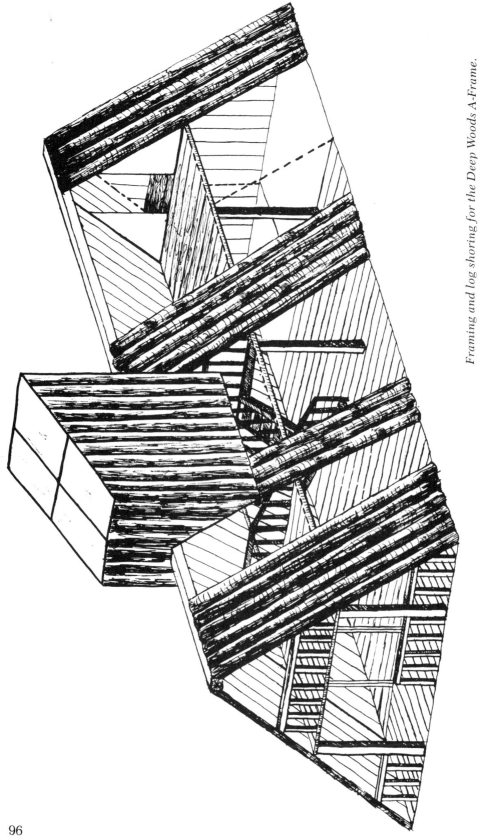

Framing and log shoring for the Deep Woods A-Frame.

5. Cut the ridge plate members. These can be any length that you can handle easily. The tricky part is positioning the ridge plate and enough supporting side logs to stabilize the building initially. The ridge plate should be at least ten inches by ten inches. Taper the ends of the shoring logs so they are flush against the ridge plate.
6. After the girders that frame the second level have been fitted to the outside shoring logs, the decking for the second level can be added, giving a working platform to aid with setting the remaining shoring logs.
7. Lay up the rear wall by setting two posts, equally spaced, between the outside edges of the floor. These two posts should extend up to the two shoring logs that form the last "A" for the roof. Anchor these two posts to the two shoring logs by using strap-iron brackets. Stack the back shoring logs against the posts and the outside of the A-frame shoring logs.

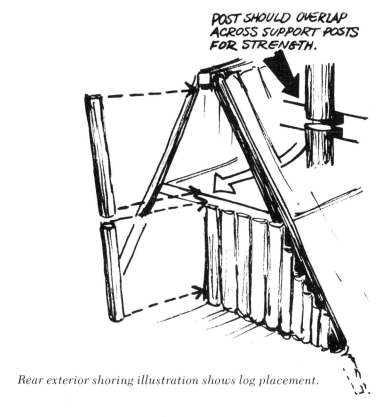

Rear exterior shoring illustration shows log placement.

8. Waterproof and backfill, using the same polyethylene, gravel and fill method used on the Trench House and Railroad Tie House designs.
9. Anchor the tops of the two back retainer posts by using deadmen anchors. Use half-inch rod that is about twenty feet in length. Insert the threaded ends through the back vertical stay posts near their tops. A strap-metal wrap at the point where the rod goes through each post will keep the post from splitting if it is pushed inward against the anchor rod and nut. Coat the rods with liquid butyl or other long-lasting waterproofing agent. Make a slit trench back into the rear wall at the excavation for

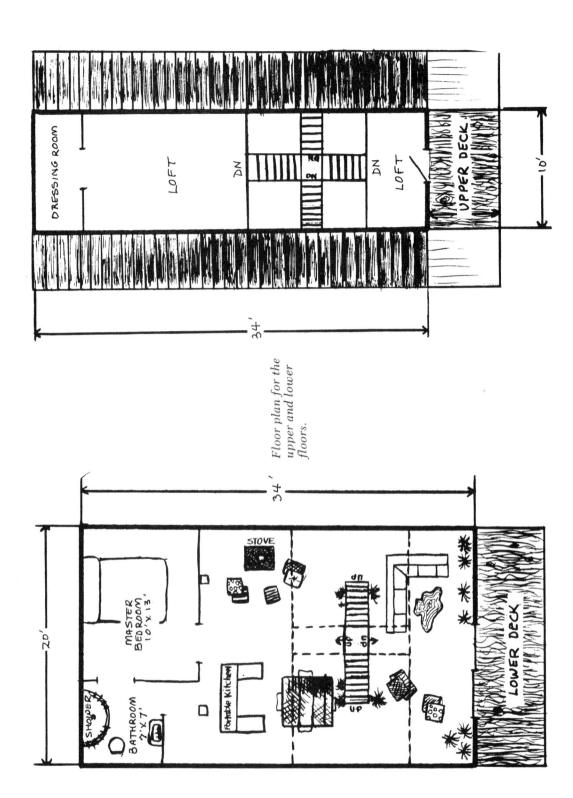

Floor plan for the upper and lower floors.

each of the twenty-foot rods. At the rear end of each slit trench, bore a post hole. Bend a hook in the end of each of the rods and cast a ball of concrete on it that will fit the post hole. After the concrete hardens, drop the ball down the post hole. The rod will then drop to the bottom of the slit trench, and the whole thing can be backfilled. You now have a simple anchoring system that will stop any kind of movement or earth thrust against the structure. If you want even more strength, just add more deadmen anchors.
10. This design can be built with or without a skylight. The skylight complicates construction, but it also permits a lot of light to enter the structure. A separate support structure will have to be built for the skylight. A square frame with four corner posts will support the weight of the skylight. These corner posts should be directly over the second-floor supports.
11. An alternative to building the skylight for this design would involve not backfilling the peak at the rear of the structure. The top five feet or so could be comprised of triple-paned glass. The peak would be fairly well protected from the wind due to the steep terrain, and the glass would break most of the heat loss. This five feet or more of glass would light the back portion of the A-Frame and help balance the interior illumination. The stairs and the opening would then be moved to the rear of the structure so that the stairwell would also allow light from the peak window to penetrate the bottom level. Sod could still be placed on the roof, but it would slough off at a five-foot distance from the peak and blend with the larger backfill and terrain. A retainer would have to be fastened to the roof to keep the sod from eroding or sliding over the edge.

Window at deck level at the back of the house in place of the skylight is shown above.

12. This twenty- by thirty-four-foot structure gives over one thousand square feet of usable space. This is a good design for heavily wooded and mountainous areas not only because indigenous materials can be

utilized, but since you do not want to disturb the environment any more than possible, this design gives the most strength and size possible in a limited excavation.
13. Approximately eight shoring logs of twenty-two-foot lengths will be needed for this design, along with about twenty internal support posts that are fourteen feet in length. Miscellaneous posts and shore logs will be needed as per your own design modifications.
14. Landscaping should be kept as natural as possible. In heavy timber country, this means not clearing any more trees than is necessary.

10
Mobile Home Earth Shelters

The tremendous rise in housing costs has brought the mobile home into the realm of respectability and economic feasibility for a large segment of our population. For those who have not lived in a mobile home, I will say that there are real advantages as well as some obvious disadvantages to this form of shelter.

The first advantage to the mobile home is its cost. I doubt that I would buy a new one because of the high depreciation of such homes. A good used mobile home can be a tremendous bargain and provide nearly as much usable space as many homes selling for six to ten times the cost. Due to the increasing costs of buying a home, builders and designers may construct more compact, efficient and economic homes. The mobile home in the interim may be a good investment if coupled with an earth shelter. Some additional advantages to the mobile home are: efficient use of space, portability, easy ventilation, good light distribution and limited maintenance. When considering the drawbacks of a mobile home, consider the fact that the units are manufactured from light materials; therefore, they are easily damaged, are usually poorly insulated and are easily damaged in wind storms. The accompanying earth shelter design eliminates the major portion of the drawbacks. The design will have to be modified for individual mobile homes, but that should not be difficult, since the design is built in stages. This design is one that will work best with mobile homes that are fourteen feet or less in width. A sixteen-foot-wide unit would be at the outside of this shelter's design limitations.

Double Envelope Mobile Home elevation.

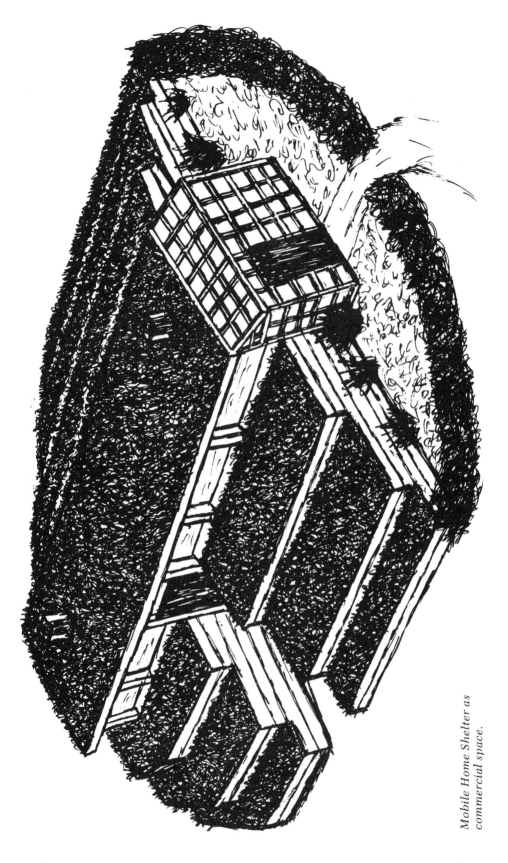

Mobile Home Shelter as commercial space.

As with any of these designs, getting land will be the real problem. If you have no desire to own land or be on an acreage, you may perhaps be able to persuade the mobile-home park owner into letting you experiment with the shelter on the park ground, perhaps with a lower rent. Whether or not you get a break in lot rent, the shelter will pay for itself in utility savings. The increased protection from wind and hail damage should reduce your insurance premiums. Sound-proofing and privacy are great side benefits, and the ability to garden on the shelter's terraces is another option.

Most mobile homes are skirted with a thin sheeting of some sort that breaks the wind and allows the insulated water pipes to escape all but the worst freezing conditions. The earth shelter will eliminate the need for skirting and make your home a visual delight. As stated, this design is subject to your modifications. I would keep as much earth mass as possible around the mobile home, though, since it is earth mass that provides the protection. The end result of combining the mobile home with an earth shelter is a double-envelope home that is probably the most energy-efficient design possible.

Double Envelope Mobile Home eliminates traditional skirting and freezing while protecting the mobile home from wind.

Construction steps

1. This shelter is specifically designed to be built on flat ground or on ground that will accommodate multiple units. It is set up to blend into the next unit; consequently, an entire trailer park could be built into eye-appealing shelters that would protect one another and contribute to the value of the property. The park owner could charge extra for these lots since the renters would realize dividends beyond the other open parks.
2. Since this design is basically another post, girder and rafter system, the corner posts have to be set first. I would make the shelter at least one foot wider than the mobile home involved. This will allow the home to be removed and another put in its place at a later date. After the corner posts are set, line up the other posts to strings drawn to the outside of the corner posts. The distance between the other posts will be determined by the type of shoring to be used and the size of the girders that will support the roof. It is my suggestion to use recycled railroad ties for the shoring and roof planking. The mobile home will act as the liner that removes these exposed ties from contact, so they become even more efficient in this design.

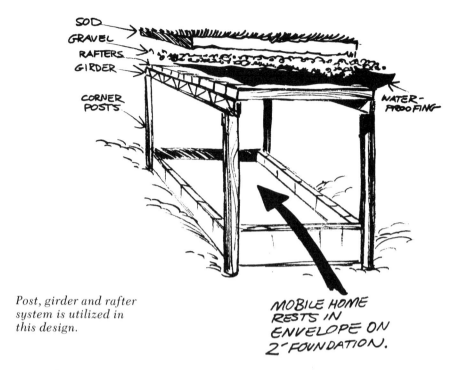

Post, girder and rafter system is utilized in this design.

3. Due to the span, yard ties will not be strong enough, so you will have to use bridge supports. Many wooden railroad bridges are being replaced with steel ones, and the wood bridge girders are available for salvage. Ten bridge girders, twelve by twenty inches in girth and twenty feet in length, will provide a two-foot overhang all the way around for a sixteen-foot-wide home. These girders placed on eight-foot centers will support the railroad ties that serve as rafter planks. A girder this size should support the rafter planks and one foot of sod. I would still incorporate a two-inch-thick gravel drain field.

Multiple Double Envelope Mobile Homes would make an attractive housing development as well as increase the mobile home's energy efficiency.

4. Notch the bridge girders and set them on the posts. Fourteen-foot yard ties will make good posts. I would use strap-iron brackets to fasten the girders into place. Floor trusses could also be used to span sixteen feet.

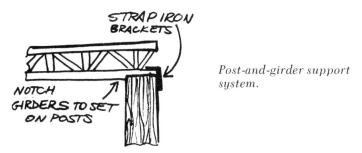

Post-and-girder support system.

5. Place the shoring ties for the first terrace. This terrace will be the widest, with each successive terrace being narrower than the previous one. The width of the first terrace will have to be somewhat in accordance with the height of the intended berm. I would recommend that the average berm height be six feet. This takes into consideration the fact that the mobile home is sitting on blocks and is already two feet off the ground. The inside roof height can be as great as twelve feet for many mobile homes. This leaves a five-foot clerestory window arrangement around the entire mobile home. This glass will add heat through the sun's radiation in the winter; in the summer it will help keep the cool interior from heating up. The end of the structure that the mobile home is

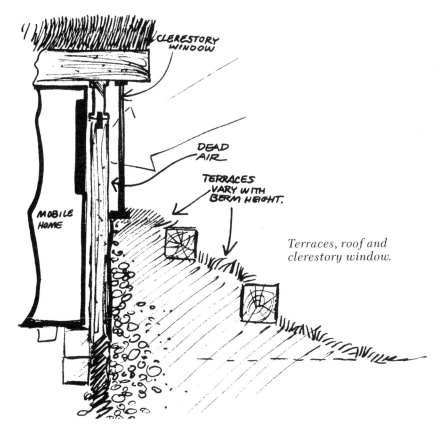

Terraces, roof and clerestory window.

backed into can be left without a berm and glassed in if you prefer. If possible, try not to make it face a westerly or northerly direction.
6. Backfill the first terrace. Each terrace should be about one foot (or one tie) deep. The width between each successive rise can be set according to the lot width. Two terraces can be backfilled at one time. Allow this soil to settle before beginning another terrace. Handle the waterproofing and drain fields the same way as for the Railroad Tie House and the Trench House. If the ties are protected by polyethylene and drained by a French drain field, they will last indefinitely.
7. Continue the shoring and backfilling to the top of the berm. This will provide an incline on which to move sod up to the roof. Be sure that openings are left in the front and back of the terrace-berms to allow egress out the back and front doors. One of the problems with the design is that it will have to be altered for each mobile home that it shelters. This should not be too great a problem with a reasonably efficient setup crew. A small rubber-tired loader, such as a "Bobcat" will speed the job up, and the ties will come out easily.
8. Place the railroad-tie, roof plank-rafters next. Waterproof this the same way that the sides are handled. A shallow slope to one side of the roof system would aid drainage and avoid the pooling of water under the sod. A two-inch drop would be more than adequate for a fourteen-foot span. A retainer rim can be fastened to the edges of the roof by using railroad ties. I would put polyethylene down over the edges of the roof to protect the rafter-planks. Float the retainer ties on the roof edge through small deadmen anchors into the roof sod. The two inches of gravel should drain the roof easily, and the floating retainer ties will let the water escape under them. The runoff will not be nearly as great as on a conventional roof, since the sod will use most of the moisture that is received in a rain shower, and the gravel will handle heavy rains.
9. If several units are to be built in this fashion, some small modifications in the berming technique will not only make the system more attractive, but will also provide common-use areas with privacy and functional use. Patios and garden ponds with tropical fish can be features that will be enjoyed by two families who face onto a common space.
10. One thing that I should have included with the first layer of terrace is the soil-pipe ventilation system. Utilizing this berming method of earth-sheltering eliminates the need for trenching when placing the soil pipe. Two or three pipes may be laid parallel to each other, running the length of the structure and covered over with the first terrace level. It is thus easier to build a plenum in which to run the pipe. By attaching an outlet from the plenum to the mobile home's furnace plenum, the furnace fan can distribute the dehumidified cool air throughout the interior. The mobile home still needs solar chimneys so that the hot air from the ceiling area can have an outlet to the outside. It is important to create openings in the earth-sheltered roof to accommodate the solar chimneys. Since the railroad-tie, rafter-planks are easily removed, these openings can be created without any major work.
11. The sewer and water lines, along with other utilities, are usually located directly under the mobile home, so this type of structure should not interfere with normal lot use in an established mobile-home park.

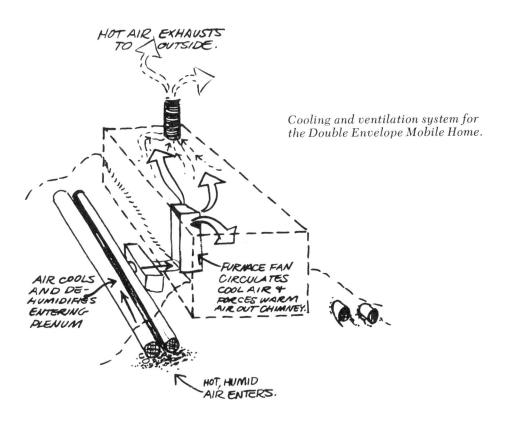

Cooling and ventilation system for the Double Envelope Mobile Home.

Landscaping can be as simple or as elaborate as desired.

12. If the bermed structure is to be built on private land, you have even more flexibility. I would build the berms wider, in this case, and have a greater depth to each level of the berm. Since it would most likely be a permanent structure for the one mobile home, a more elaborate terrace arrangement could be established. The terrace levels could match the mobile home's individual window level for the greatest amount of light. The entry and exit walkways could run in several directions, giving access to specially created levels. These levels could include an eating level, a sitting level that's complete with a fish pond, a level for croquet or badminton, and even a level for just watching the stars. Your imagination would be the only limit imposed in the creation of this structure. Small fruit trees could be established in an orchard terrace. A garden level and a swimming pool could be created. The utilization of a large terrace would cause the swimming pool to disappear into the ground without your having to excavate.
13. Approximately 328 ties would be used in the creation of a structure for a fourteen-by-eighty-foot mobile home. The cost of building your home could be reduced greatly through good bartering or volunteer labor. The bridge girders will be the single most expensive units.

II
Round Concrete Structures

I have created three designs (the Daylight Dome, the Easy Dome and the Freedom Tunnel) that use poured concrete almost exclusively in their construction. These three designs are round in form, which is not a necessary shape for using concrete underground, but it is a shape that is very strong and easily constructed by the amateur or self-builder.

Daylight Dome

Although it is one of my earlier designs, the Daylight Dome still seems to be one of the most popular designs. On a square-foot basis, it would cost ten to twenty percent less to build than a conventional above-surface structure of equal size, depending on how much work you can do yourself.

The circular shell can be constructed out of cement blocks or poured concrete. Since the shape is round, the need for buttress walls or other internal support systems is eliminated. If I were to construct the shell out of cement blocks, I would run a bond-beam through the center course and one as the top course. Since this structure is comprised of two levels, the height lends itself to a hill cut more than a flat land-berm. About one-third of the structural shell is glass that forms a south-facing greenhouse. This greenhouse supplies the greater part of the structure's winter heat needs. The chimney-shaped skylight in the dome supplies some additional winter heat while balancing the interior light.

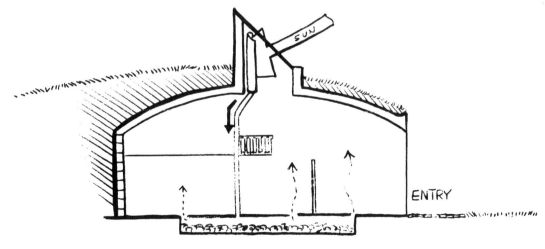

Cutaway of the Daylight Dome illustrates solar gain through the light well and the heat storage area located in the rock area under the floor.

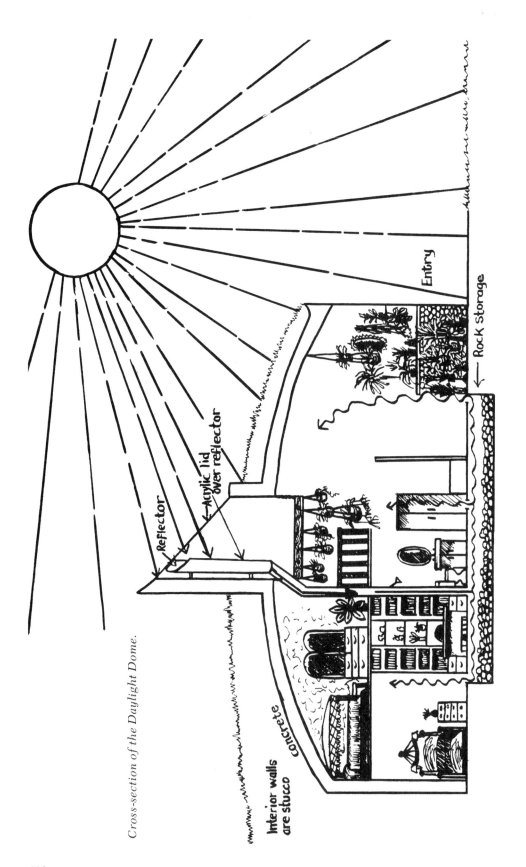

Cross-section of the Daylight Dome.

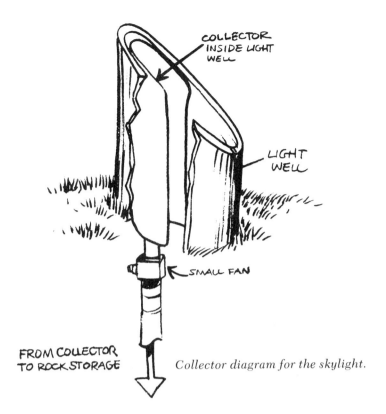

Collector diagram for the skylight.

Use of cement blocks

You won't need to hire firms to set up for the shell's walls if you use cement blocks. The framing for the glass is a simple carpentry job. Pilastre posts at the ends of the cement-block walls will help keep them from moving when they come under compression from the backfill. If the walls move even a small amount, it would cause the greenhouse glass to shift, crack and break. The cement blocks would be surface-bonded instead of laid with mortar. The cores on six-foot centers should be filled with concrete and four bars of half-inch reinforcing steel. These columns would provide more shear strength and help keep the wall from moving.

Roofing

The roof could be formed over a series of scaffolding and plywood sheets, but an easier method would be to mound-form the concrete. This method would entail filling the shell with fill soil or sand to the top of the walls, then mounding the sand into the shape of the dome. At that time the reinforcing steel could be placed and the concrete poured. After the dome is set, use a loader and remove the fill.*

Light well

The dome would be connected to the walls through the reinforcing rods left protruding from the pilastre columns of filled block cores in the wall. The dome is

---

*If the structure is being built on land in a rural area and a driveway needs to be constructed, then most of the fill that supports the dome-pour could be composed of crushed rock. The rock could then be put on the driveway after removing it from the inside of the shell. Only the actual dome would be shaped from the fill sand.

further reinforced by the light well that is part of the dome top. This light well should be formed and reinforced with steel rod, that is in turn tied to the rod reinforcing of the main part of the dome. The light well then becomes a round flying beam that adds extra support to the dome.

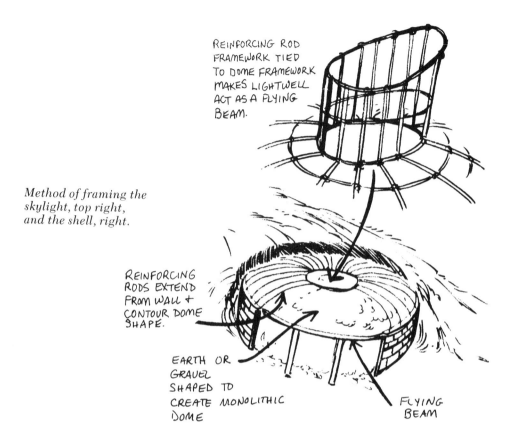

Method of framing the skylight, top right, and the shell, right.

Greenhouse framing

The portion of the dome that spans the greenhouse framing will have to be built into another flying beam in order to be self-supporting at that area. The greenhouse glass framing is a nonsupporting unit and, therefore, the entire dome has to support itself. Thickening the edge of the dome over the greenhouse and adding additional reinforcing bar will take care of the problem. Here again, the exact incidence of the dome's curve, its poured thickness, and the number of reinforcing bars to be used will have to be determined by a qualified engineer. The engineer's figures will be based on the diameter of the structure and the curve needed to support the imposed weight over that area.

Insulation

I would spray the exterior of the dome and the walls with polyurethane foam insulation. Depending on the local climate, the foam would extend down the entire side walls or down just about four or five feet. Generally, the longer the heating season, the more insulation is needed.

Even in extreme hot climates that do not experience frost, insulation is needed on a concrete structure to keep the heated earth above from transferring that heat into the structure via thermal conductivity. Determine at which point the earth temperature stabilizes in a hot climate, and end the insulation at that point on the walls. Below that point, the walls will conduct the coolness of the surrounding soil to the interior.

In a very cold climate I would spray on four inches of polyurethane over the dome, and reduce the thickness to three inches on the upper wall and one inch on the lower wall. If the floor is going to be poured, both it and the bottom of the footings should have at least half an inch of insulation in order to stop the heat from bleeding into the soil.

Construction tips

- ◆ The waterproofing would be done the same way as is done for the structures previously described. A good gravel drain-field, connected to drain tile that exits away from the structure, and a polyethylene sheeting will easily protect the sprayed polyurethane insulation and walls.
- ◆ One additional thought about backfill: Since I have recommended filling the shell with rock and sand to form the dome, the walls will have to be stabilized during this operation or they might otherwise be pushed outward by the fill inside. I would recommend doing the backfill as the rock fill is added inside the shell, stabilizing the walls by providing equal pressure. Once the dome is poured and set, it will give the walls the extra support they need when the inside fill is removed.

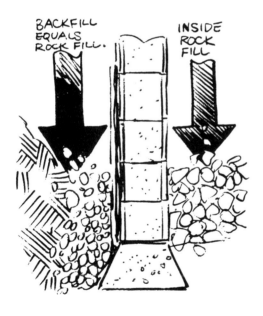

Block wall, bell footing, exterior rock drain field and waterproofing are shown, along with temporary rock fill, forming and supporting the roof of the Daylight Dome.

- ◆ Be sure that the support attachments for the second-level deck are secured as the wall is built. As the drawings indicate, the back inside surface of the light well has a dry plate collector to increase the sun's radiation in the winter. A small fan and duct are fastened to the dry plate collector in order to transfer heat into the insulated rock storage below the first-floor level. The extra mass storage will supply heat for the evenings and on cloudy days.
- ◆ The floor area of the greenhouse is composed of soil which is laid over with loose brick. This makes the greenhouse floor accessible for planting.

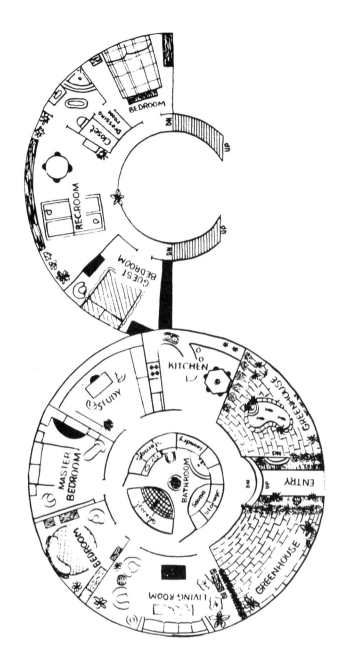

Daylight Dome 1½ floor plan.

The interior

Since this is a clear-span structure, the inside can be finished off just about any way you please. The suggested floor plan depicted in the drawings has appeal to those who prefer an informal life-style. The bathroom in this design becomes the central focus of the structure. The way it is planned, the bath is more of a comfort center than a bathroom as we think of it. A hot tub could be substituted for the sauna if desired. The light well illuminates the bath and provides an almost tropical setting with potted and hanging plants. The plants will thrive very well in the light and humidity. The floor-to-ceiling storage that is built entirely around the central bath core takes care of the majority of storage requirements for the structure and frees more room space for use.

The two bedrooms are the only rooms that I would partition off with solid walls, while the other areas would be separated by plants, trees and movable hedges. By varying the levels of plants through the interior, visual monotony is avoided.

Additional construction information

- ◆ The dry-stacked block wall can be constructed by an amateur without difficulty.
- ◆ Trowelling on the surface-bonding is simple, and the fill for the dome involves using a front-end loader.
- ◆ Placing the steel reinforcing rods for the dome may require some help from an experienced hand, but that is about the only technical job for which you may need assistance.
- ◆ The sprayed polyurethane can be contracted at fairly competitive rates.
- ◆ The footings for this structure should be bell-shaped and wide enough at the bottom to accommodate the weight of this clear-span building. All of the weight of the wall and roof will be transferred to these footings. Again, get sound engineering advice in this area. The footings will have to be specified according to the type of soil on which the building will rest and the total load of the building. This particular shape of building withstands expansive soil pressure very well and can be placed in a deep-hill cut without too much concern.

Easy Dome

This next design is also a dome, but in its purest form. It is a single-level dome that can be built in almost any size, although a thirty-foot diameter is an easy, workable size. I call this the Easy Dome because of the ease with which it can be built. It would be about the most economical of the concrete designs to build since it is almost a single-medium design—few additional materials and labor will be needed.

Substantial footing will be required since this is a clear-span building with no internal support to spread the footing load. After acquiring the engineering figures, pour a bell-shaped footing with reinforcing bar stubs protruding from the of the footing so that the poured dome becomes monolithic with the footing. This design can be placed just about anywhere, but it would blend very well with the environment if placed in a gently rolling or hilly area (the sandhills of western Nebraska, Kansas and the Dakotas would be the ideal terrain for this dome). Gentle mounding of the backfill on the dome would make it almost indistinguishable from the surrounding topography. Rather than make a hill-cut in the rolling terrain, I would build it between two hills, making it a smaller hill that blended with the other two, so that some of the surface of the other two hills could be pulled around as backfill.

Easy Dome elevation causes it to blend well into the surrounding terrain.

Forming the shell

Decide on the size of dome that you need, pour the footings, and then use a front-end loader to mound the dome's form inside the footings. Once the dome shape is completed, you are ready to place reinforcing steel. Once again, get qualified engineering specifications for the size of dome you wish to shape. After the shape is perfected and all of the steel is in place, begin the pour. You may need a cement pumper to raise the concrete to the top of the dome.

The thickness of the shell will be determined by the shape of the actual dome and the use of light slots. A dome with the right shape will distribute stress

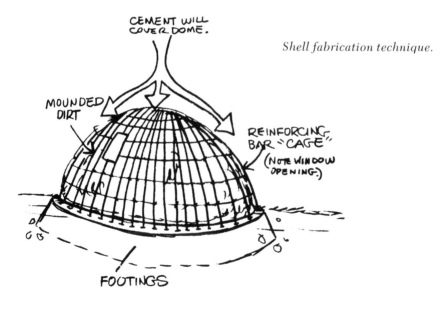

Shell fabrication technique.

equally over the shell's surface, thus requiring less reinforcing and a thinner shell thickness. You can opt for more steel and a thicker shell wall, but the cost will be greater, so I would spend the money for the engineering and save on the materials.

One of the drawbacks of the American Indian earth lodge was its lack of natural light. A similar problem occurs with this dome. I have designed elevated light slots. These slots could be lowered to become south-facing windows, but backfill retainment then becomes more difficult. The backfill must be spread over a large base area in order to support the upper dome cover. The top half of the dome will need only about a foot of earth if the dome is properly insulated. I would spray on a cover of polyurethane and apply a layer of liquid butyl over it.

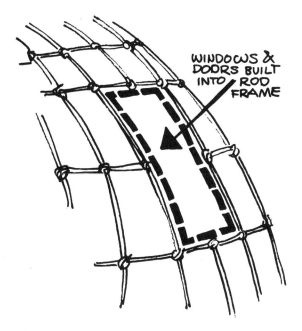

Window and door placement.

The drainage would be handled the same as for all of the other structures in this book. A gravel drain field tied to drain tile that leads away from the structure is integral to every sound underground design. Proper drainage means that there will never be any worries that the shell will leak.

A liquid sealant that sets up but does not harden is essential to waterproofing a compound curve such as a dome. Polyethylene sheets would bunch up, and water consequently would tend to capillary under overlapping seams. Be sure that all of the roof openings needed are cast at the time of pouring, or it will be difficult to go back and cut openings later. Sewer stack, chimney, skylight openings and all exit doors will have to be planned into the stress calculations.

Insulation

Insulation values for this and all underground structures should follow the guidelines set forth in the Daylight Dome design. Insulate according to the length of the heating season. The coldest climates require a complete cocoon, while hot climates require insulation on the roof and side walls only to a depth where the soil temperature stabilizes.

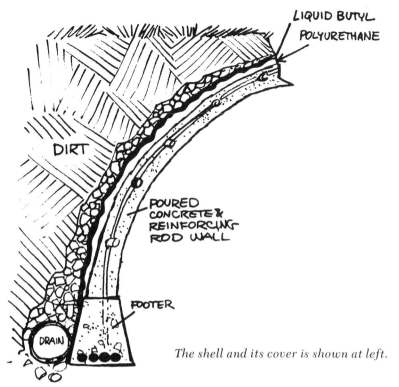

The shell and its cover is shown at left.

Skylights

Skylights are designed to serve as solar chimneys and vents, as well as for light distribution. The small ones at the highest elevation on the dome open for ventilation. These windows will heat up the summer and when opened during the hottest period of the day, will draw the hot air up and out of the interior.

Ventilation

Soil pipe should be used in this structure (as it should be in every underground design). Pipes for the dome can be placed on concentric circles around the base, thus taking up less room than the lateral type. Four separate pipes with two inlets each into the structure will supply all the fresh, cool, dehumidified air the structure will need. The inlets for each pipe should enter the dome at a different location so the entire structure receives an even distribution. Plan these inlets so that at least one goes into any permanently divided room.

Permanent or solid partitions should not extend all the way to the dome. Curve the dividing walls to follow the dome's line, and note that a seven-foot maximum wall height is plenty for privacy. This allows the air to circulate naturally throughout the structure. The hot air in the winter will go over the dividing walls, forcing the cooler air under raised doors and cold air vents cut at the bottom of these dividers. During winter months, the skylights will create enough heat for the interior during the day. This warm air will be forced to the back portion of the structure and in turn push the cooler air back to the front of the interior to be warmed and cycled again. The soil-pipe inlets should be dampered so that they are closed off in the winter; however, one or two soil pipe inlets should be slightly opened to allow fresh air into the interior. The soil pipe not only cools and dehumidifies the air in summer, but also tempers the air in the winter so that it

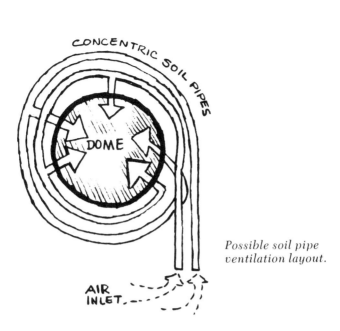

Possible soil pipe ventilation layout.

does not enter the interior at its ambient temperature. The temperature of the soil pipes should remain between fifty-five and sixty-five degrees Fahrenheit, depending on constant soil temperatures below frost or heat lines.

The interior

Since the shell of this design is monolithic, the ceiling and walls are part of a continuous form. This means that a single surface coating prepares the entire interior and, therefore, should save money. I am strongly in favor of dividing interior space as naturally as possible and avoiding fixed partitions as much as possible. Excavated conversation pits, raised study areas and built-up plant areas will provide visual stimulation and help divide the interior naturally. Bedrooms and baths are the only areas that need fixed partitions. The utilization of open space not only gives more usable space in a dome and avoids pie-shaped rooms, but also promotes better light distribution and balance. A nine-hundred square-foot space takes on the appearance of twice that size when left open.

Multi-dome complex

Expansion of the Easy Dome structure can be achieved by connecting a series of smaller domes through tunnels. The expanded structure system would make an excellent school, with each pod being an individual classroom; proper utilization of plants and shrubs can result in the creation of several classrooms. Due to the ease of construction, these pods could be constructed in almost any location that is within transportable distance of ready-mixed concrete. If the project warranted, a mixer could be set up that would mix enough cement for a continuous pour. The shell has to be done in a continuous pour or it will fail structurally. There can be only one cold seam, and that is at the footing base.

Freedom Tunnel

I call this design the Freedom Tunnel because it will free you from high utility bills and house payments and because it is tunnel-shaped. It is formed by the same construction methods used to build the Easy Dome. The dome and arch are

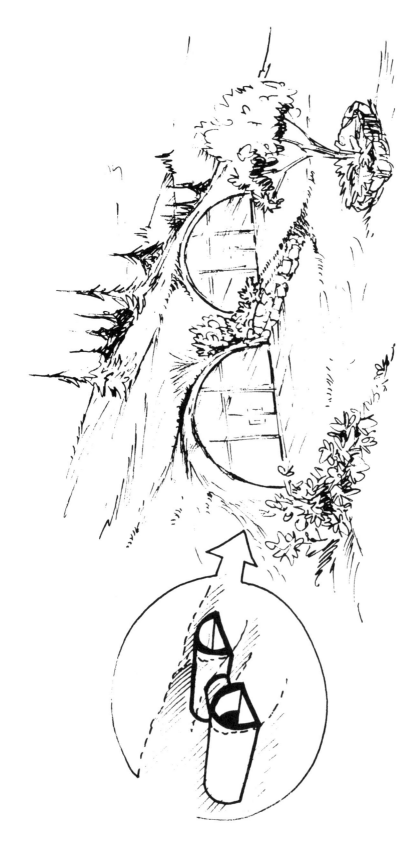

Elevation of the Freedom Tunnel depicts its simple elegance.

good shapes for underground construction since both distribute the imposed weight of the soil over the entire surface of the structure. A pair of these tunnels set in the ground parallel to each other would provide a reasonable amount of usable space for two or three people. The recommended size for each tunnel is sixteen feet wide by thirty feet deep.

The Freedom Tunnel can be formed by the amateur self-builder without forms or scaffolding. Flexibility of the tunnel within various design concepts is probably greater than any other shape. They work well in both flat or hilly terrain. Schools, offices, hospitals and dwellings can assume almost any configuration—from spokes of a wheel with a central courtyard to stacked multiple-story uses.

As the illustration indicates, each tunnel would be connected by a corridor tunnel that would be cast monolithically with the two main units. The light wells are cast with each unit so that there are no cold joints in either unit. The units and connecting tunnel are formed by mounding up fill into the desired shapes. The cost of construction can be greatly reduced if you have access to free fill from a county road or bridge project or if you can use the soil from the excavated building site. Good engineering specifications will further reduce the material costs by properly configuring the arch to use the least amount of concrete and steel. Bell-shaped footings should be cast with a kicker plate on the outside to keep the arch from pushing outward when loaded. The light shafts provide auxiliary heat through a collector plate that is mounted in each well. The heat is drawn off through ducting into a rock bed below the floor of each unit. This rock storage heats the interior of the units at night or on cloudy days. The glass fronts of each unit create heat and distribute light to the interiors.

Pouring concrete

Concrete will probably have to be pumped to the top of each tunnel and then distributed downward by hand. A properly configured four-inch shell will adequately support evenly distributed backfill of any desired amount. Have everything set up so that once the pour is begun, it can be continued until completed. This might be a good time to get some volunteer labor for a few hours. Concrete gets very heavy in a few minutes, and that is why it is pumped to the top and pushed down. Be sure the mud is stiff enough to prevent it from running down and piling up at the base of the walls. Thick mud is really stronger when cured than when it is overly wet cement. It also takes too long to cure out when the mix is too wet.

Waterproofing

Waterproofing and drainage for this design should be handled in the same manner as called for in previous designs. As I have mentioned, I do not intend to create submarines out of any underground design. Drainage and water management are far more critical than an expensive external waterproofing agent. A raincoat that handles an excess of water on a temporary basis basis is much more economic and just as useful. The combination of polyethylene and a gravel field is just as effective as some of the exotic rubber sheeting and plastic coatings on the market—at a fraction of the cost. These have their place when the more economic method will not work (but most of the time it will). The polyethylene sheeting will not work well without bunching and risking water infiltration on compound curves, but it will work well on the shells of the Freedom Tunnel.

Insulation

Since the shell is composed of concrete, it will have to be insulated. Here again, I would spray on polyurethane foam. It seals around curves where rigid insulation will not. Almost any plastic sheeting or foam insulation, whether rigid or liquid, will degrade from ultraviolet light, so the insulation must be below grade or otherwise protected.

Structures that are constructed from noninsulating materials such as concrete, steel, fibreglass and relatively thin sheets of wood shoring all need exterior insulation to keep out the cold in the winter and heat in the summer. Since the roof and upper sidewalls of any structure are the most vulnerable, these need the most insulation. The lower walls, floors and footings only need insulation in very cold climates.

The interior

All clear-span structures present the greatest amount of interior design use and possibilities, and the Freedom Tunnel is no exception. The floor plan shown is only a suggestion of space usage, and you can use it as a basis for developing a plan appropriate for your own needs. The floor plans that are included in this book are designed for the most efficient use of the space involved and will meet the egress requirements of the building codes in most areas. Consider the following when planning the interior of your structure:

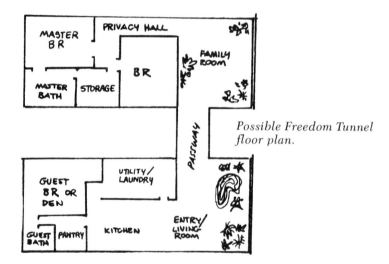

Possible Freedom Tunnel floor plan.

- ◆ Use plants and shrubs as space dividers and elevation change accessories. Plants absorb sound and replenish the oxygen.
- ◆ Removable and easily moved partitions allow you to accommodate the changing needs of your family without great effort or expense. I have personally experienced a living environment of plants, water and fish in my own home and would not go back to sterile walls and cubicles out of choice again!
- ◆ The interior shell surface can be finished by coating it with drywall compound that is troweled on like plaster. Paul Isaacson, a fan of round houses, coated the inside of his curved walls with drywall compound and then

sanded it smooth. Everyone who came to his home could not figure how he had curved the drywall sheets so smoothly to fit the wall surface!

◆ Drywall compound used to cover walls can be painted. I like to paint everything white that is a divider or wall surface so that I do not have a problem with furniture colors, drapes, wall hangings or paintings clashing with surface color. The white surfaces will multiply the light back to you, as well.

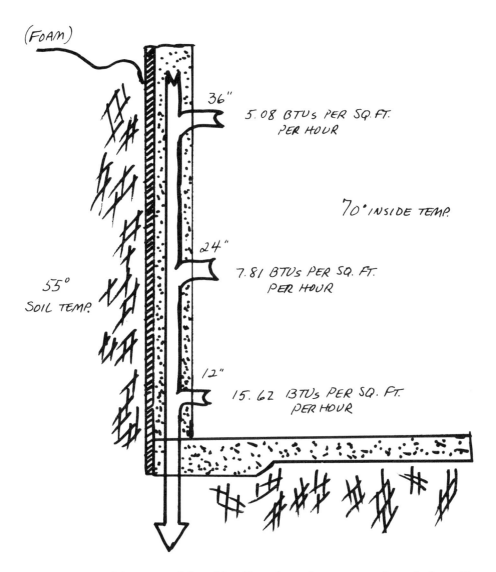

A non-insulated footing and floor bleed heat from the structure through the wall.

The wall shown above is made of concrete and has foam insulation on its exterior; the floor is concrete and is in direct contact with the soil. The illustration shows how much heat is siphoned out of the wall and through the floor

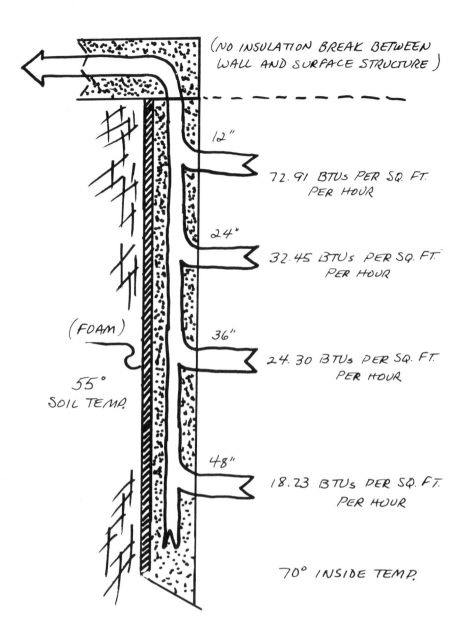

The lack of insulation between the underground structure and the surface-exposed structure causes heat loss through the wall.

when the footing and floor are not insulated. All BTUs are based on a per-square-foot measurement of surface exposure for one hour. Insulation is needed under footing and floor in areas having cold climates and a long heating season. In a hot or tropical climate, no insulation in the lower wall, footing or floor would be needed.

The illustration on this page shows an underground wall with foam insulation on the exterior surface, with no insulation (such as an earth retainer wall) between it and the exposed surface structure. When there is no insulation break, heat loss from the entire structure, emanating through the siphon on the walls, results. In hot climates, the opposite would be true and the underground structure would gain heat from the surface structure into the walls. An insulation break is always necessary between underground structures and exposed surface structures or appendages.

Climatic map.

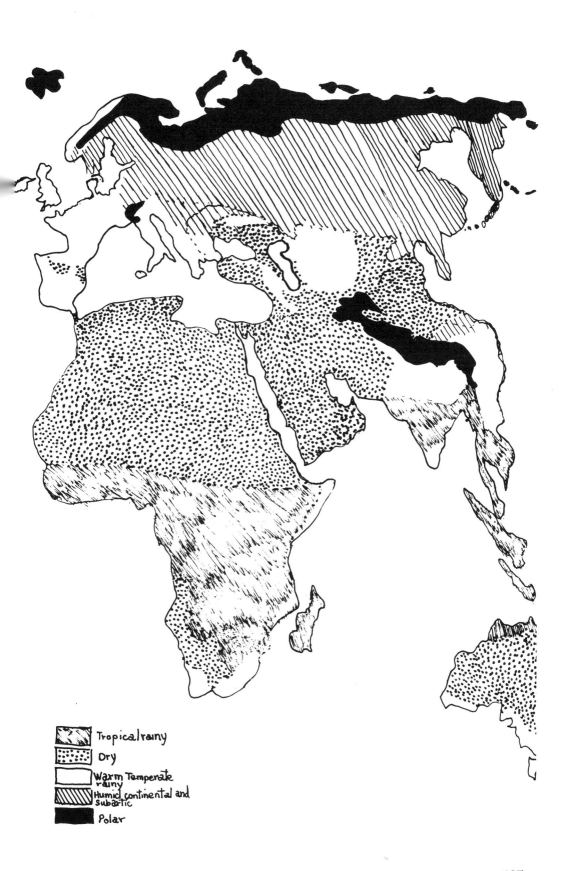

127

12
Convertible Crescent

This design is truly convertible. The main structure can be left entirely open, the exception being the dividing wall between the garage and the living area. What makes this possible are foam-laminated walls, which have a four-inch polyurethane core. These lightweight dividers are mounted on hidden tracks in the floor and ceiling and serve as excellent sound barriers. You can have as many bedrooms or usage areas as you wish by simply multiplying the number of dividers throughout the space. When a totally open-use area is required for a large party, just move the dividers to one end of the structure. You can accommodate a large number of people in a huge hall for a cocktail party, or have any number of friends over for the night or weekend and provide private bedrooms for each of them!

The greenhouse across the front provides heat and light. The six-foot-wide hall at the rear provides a fire route out of the building. It also serves as a storage area and a place to move furniture while expanding the open space when moving the dividers.

Wall construction

A central plumbing core reduces expenses and makes maintenance less frustrating. The design of this house makes it fairly simple for the self-builder to construct. The curved back wall needs no extra internal bracing and only minimal steel reinforcing. This design can be constructed by pouring the three main walls or by using cement blocks with surface-bonding. If blocks are used, be sure to fill the cores on six-foot centers with cement and rebar. A bond beam will be needed to serve as the top course of blocks to tie the wall to the poured cores, thus securing it against lateral and vertical expansive forces.

Constructing the roof

The roof can be poured as a monolithic structure complete with its own joists by using earth fill inside the shell as support, or a truss system (using kick-out 2 by 4s and four-by-eight sheets of half-inch plywood). You are going to need 2 by 4s for greenhouse framing, so some of the pouring supports can be used later for framing. The remainder of the 2 by 4s and plywood should be in almost new condition and can be resold upon completion of the pour. The reason this material should be in such good condition is the fact that no nails are used in the lumber, and the surface of the plywood used to support the cement is protected with polyethylene sheeting.

The illustration below shows how this *truss system* works. Everyone will tell you that this system will not work, but it does. It is quick and requires far less material than a vertical bracing system. This system supports on four-foot squares. Basically, if the system will support a two-hundred-pound man as he walks across it, it will support the wet weight of the concrete. Joists can be formed by nailing 2 by 12s similar to an upside-down, bell-shaped footing form. This

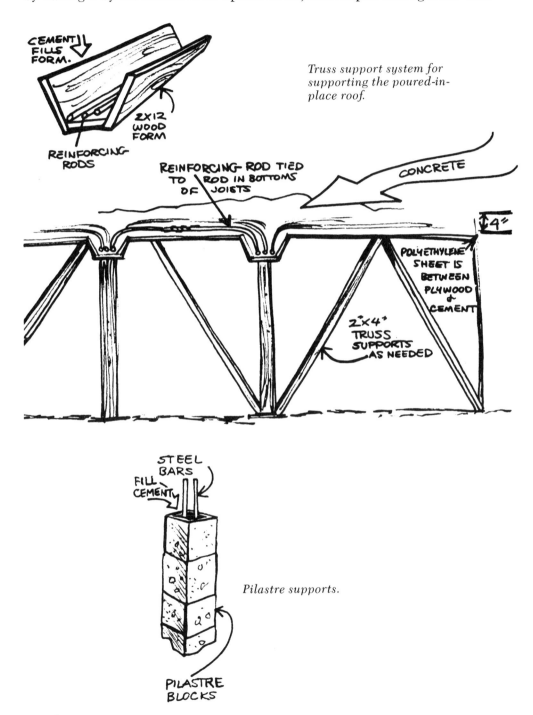

Truss support system for supporting the poured-in-place roof.

Pilastre supports.

Convertible Crescent...

makes it easy to strip the forms later when the roof has set up. This form is just as strong as a square one and uses less concrete. Place the reinforcing rods near the bottom of the joist form and tie it to the roof-reinforcing steel. The joists are on eight-foot centers, or nearly so. The curve prevents exact eight-foot centers, but they will be close enough to support the eight-inch-thick deck.

A flying beam will add extra support to the inward curve of the roof at the front. Pilastre columns under each joist at the front of the structure will complete the roof support system and provide the basic framing supports for the greenhouse glass. The pilastres can be poured or constructed from specially cast pilastre blocks. There are paper tubes that are used for pouring columns, except for the round columns which are hard to frame in. My preference would be to build the columns out of the special pilastre blocks. These blocks are very dense and have small slot-core openings. These cores should have a couple of bars of steel each, and then be poured with cement. This will form a very rigid column with good shear characteristics; compression load carrying capability will be very good.

The flying beam at the front edge of the roof will not only help support the roof, but will also act as a retainer for the earth cover it will receive later. The slanted back side of the beam will allow the heaving soil in winter to slide harmlessly up and down its surface without presenting a vertical face that would probably crack. The roof should be cast so that there is at least a four-inch drop from front to back. The top and side drain fields will carry the roof runoff down to the drain tile at the footing and away from the structure.

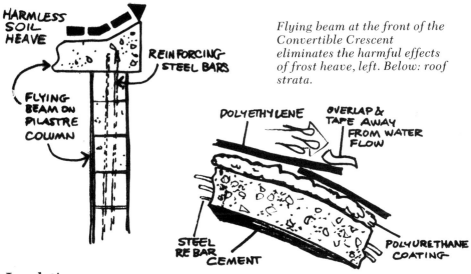

Flying beam at the front of the Convertible Crescent eliminates the harmful effects of frost heave, left. Below: roof strata.

Insulation

Since this is also a concrete fabrication, it will need insulation. The polyurethane foam sprayed on the roof and walls as needed will make the best insulation for this design because of the curve. Polyethylene sheeting over the urethane will work on this design because there are no compound curves. Overlap the sheeting, and tape the seams. Remember to overlap away from the water flow in the same manner that shingles are laid down. Polyethylene sheeting is much less expensive than the liquid butyl and other liquid sealants that are needed on such compound curves as the dome. The polyurethane will not absorb much moisture anyway, but it is a good idea to protect it to keep its insulation value intact.

13
Starlight Earth Lodge

The Starlight Earth Lodge design, perhaps the simplest of all designs discussed in this book, utilizes the crest of a hill that drops away on all sides, with inclines that make the hill all but worthless for growing anything but grass. Using land that cannot be used for crops lends further credence to this basic but beautiful habitat. Rugged terrain like this is found almost everywhere. It does not matter if the hill is not very steep; the circular shape of the structure lends itself to hilltop siting. The fact that the walls of the structure will be the stabilized excavation itself, demands that construction of the structure be done on the crown of the hill or on flat land where slope shift will not be a problem. This siting allows rapid runoff of water so that there will be no danger of water seepage into the structure.

This very simple structure has its roots in American Indian history and recent development of the geodesic dome. The earth lodges of the Mandan Indians and the assembly house used by the Miwok Indians served as models in the development of the Starlight Earth Lodge structure. In contrast, the center is covered by a modern geodesic dome that can be built from scrap or recycled material for little cost; labor to assemble it will be the primary expense.

Elevation of the Starlight Earth Lodge.

Excavation

The shell's wall is the excavated earth. The drawing below shows a cutaway view of the basic construction. The excavation can be done by hand or machine. A few willing hands and several wheelbarrows will allow this forty-four-foot diameter excavation to be dug in a couple of weeks. If machines are used, care must be taken so that side walls are not damaged; otherwise, they will be unuseable as natural walls that are plastered. My suggestion would be to use a Ditch Witch trencher to cut the perimeter to a depth of six feet. A backhoe can then be used to cut to this trench.

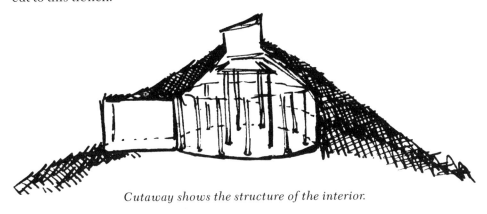

Cutaway shows the structure of the interior.

Hand-shaping and smoothing of the clay will finish the surface in preparation for plastering. Be sure to save the topsoil and sod for placement over the roof when the framing is complete. The perimeter trench should then be measured from the clay subsoil down, not from the topsoil down. Try to remove the topsoil in one-foot-thick blocks of sod for easy placement on the roof. The topsoil should be removed to a point on the crest of the hill where there is a slope that is far enough from the excavation to keep the water from accumulating.

Minor excavation creates the stabilized walls of the structure.

Placing the roof system

Four center posts, thirteen feet long and one foot in diameter, have to be cut. These can be comprised of any substantial wood, perhaps from an area where you have thinned out the trees. These should be set three feet in the ground on a flat stone. The posts should be treated with an application of liquid butyl rubber and then wrapped in nonbiodegradable garbage bags, set with a few shovels of large diameter gravel and filled with soil. The outer perimeter posts should be set in the same manner. These posts should be eleven feet in length and eight to ten inches in diameter. The tops of the posts can be notched to receive the stringers, or they can be cut, using the natural fork of the tree.

All posts, stringers and radial beams should be cut, peeled and dried at least six months prior to building. They can be treated with liberal doses of linseed oil as they dry in order to keep them from splitting. As a result of linseed application, they will have an attractive luster.

Post treatment.

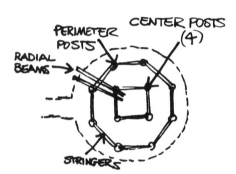

Post-and-beam layout.

The stringers should be ten to twelve inches in diameter, the radial beams about five inches in diameter, and crosstick sheeting members about three inches in diameter. This makes a tremendously strong roof system. When covered with two layers of polyethylene sheets, six-mils thick, and sod, the roof is not only weatherproofed and self-insulating, it is also pleasing to the eye.

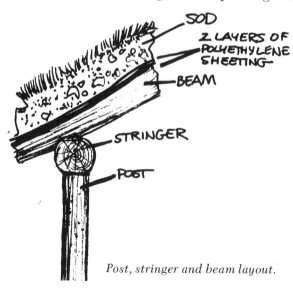

Post, stringer and beam layout.

Creating an insulation break

The secret to using the excavation itself as the perimeter wall is an insulation break that is cut just under the fill at the lower part of the earth cover. This break is a trench that is dug below the frost line around the perimeter of the backfill, using a Ditch Witch or other trencher. Two-inch-thick beadboard that is protected by garbage bags made of non-biodegradable plastic is inserted into the trench. This will form an insulation break that will keep the walls of the structure from freezing or breaking as a result of frost heave.

The inside walls are to be covered with metal lath or fine mesh chicken wire and then plastered. Three shovels of milled clay dirt, eight shovels of sand and one shovel of portland cement make up the plaster mix. The mixture solidifies and keeps its earth color.

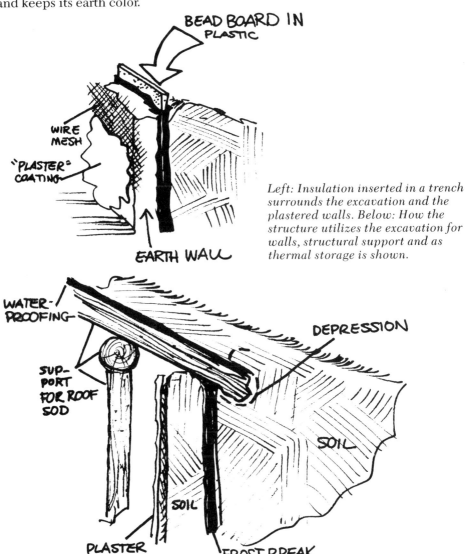

Left: Insulation inserted in a trench surrounds the excavation and the plastered walls. Below: How the structure utilizes the excavation for walls, structural support and as thermal storage is shown.

Radial roof beams

The radial roof beams will have to be set in a depression about two feet back from the rim of the excavation (which serves as a "kicker plate" to receive downward thrust of the roof weight). The depression should be lined with polyethylene. The total rise of the roof to the point where the dome sets is approximately four feet. You can now see how this will blend nicely with the crown of the hill. The roof composition consists of posts, stringers, radial beams, horizontal covering poles, mud plaster to smooth the surface, two sheets of six-mil polyethylene, and six to eight inches of sod. All pieces can be held in place until under compression by aluminum wire (hide thongs were used in the Indian lodges).

Types of center domes

The crown of the structure is a twelve-foot-diameter geodesic dome. The corners of the center square will have to be filled with poles to form the near circle that will be required to receive the dome. The dome should be set on the outside of a ridge, covered with plastic and soil so that water will not capillary back to the center opening. Contact any local engineering firm for specifications for geodesic domes. The dome can be made of scrap 2 by 4s, and the glazing can consist of scraps of quarter-inch-thick plate glass that are cut to size. If the dome is too difficult to construct, a simple A-Frame of 2 by 4s on two-foot centers, glazed with recycled storefront glass, will work fine. A pyramid would look attractive atop the house. The type of center roof glazing to be used is up to you; try to use recycled or scrap materials whenever possible.

Various coverings for the center of the Starlight Earth Lodge.

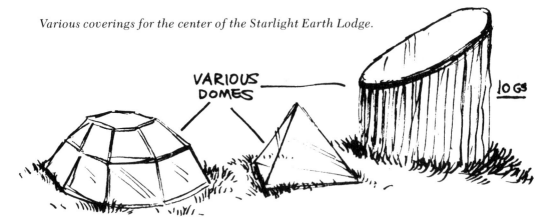

Entrance tunnel

The entry tunnel is made of posts that are five or six inches in diameter and set into the ground. They are to be bridged by the same size of poles over the top with a polyethylene-and-sod cover. Wood is self-insulating, so an insulation break between the entrance tunnel and the main structure will not be necessary. I would make the tunnel face south, and try to make it as inconspicuous as possible. The tunnel will be eight feet wide, eight feet tall, and can be glassed in on the front. It will have to be from eight to ten feet deep to allow the roof backfill to continue its contour. Not much of a facade will present itself, except for the tunnel entrance and the center dome or glazing. It is the center-glazing that makes this dwelling enjoyable and full of light. The opening can be protected by a woven Indian rug that is rolled up in the morning to allow the sunlight and heat in during the winter. The rug should be sewn to poles that support it as it is drawn across the opening on pole rails. This rug is an attractive protection and will cut heat loss through the glass during winter nights.

Entrance to the structure.

Movable insulated and decorative internal covering for the center opening.

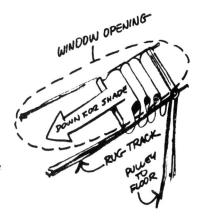

Additional comments on the floor plan

If the roof glazing has a door in it and a pull-down ladder that will extend upward to it, the roof can serve as a second escape exit. The light from this center opening will amaze you, as there will not be a dark corner in the structure. The floor plan is only a suggested one, but I like it since it keeps the space open for good air movement and use of space. I have tried to keep as much of the original cultural heritage of the structure as possible without making it uncomfortable. The heavy use of plants and bright-colored-fabric wall hangings lend both life and color to the interior and serve as attractive room dividers.

One further note: The greenhouse effect of the center glass will create a lot of heat in the winter, but a back-up system such as a rammed-earth Russian-type fireplace would work well and serve as a heat storage mass. Be sure to leave an opening for this flue, and vent the stack for the composting toilet. A rammed-earth cooking stove can be created at little or no cost. If you do not wish to cook using wood, a gas burner top can be inserted into the earth base. Summer cooling can be aided by circling the structure with twelve-inch soil pipe that is buried below the frost line and brought into the structure at six or eight locations along the outer wall near the floor. This cool, dehumidified air will be drawn in by the hot air that is rapidly escaping out of the dome through several panels that serve as thermal chimneys. The hottest time of day outside should be the coolest time of day inside because the air exchange will happen more rapidly. Four hundred feet of one-foot-diameter soil pipe that enter a common plenum in four one-hundred-foot sections provide the equivalent of three tons of air conditioning.

14
Cordwood Courtyard

This is another design that works well at the crown of a hill or as a bermed structure on flat land. Its motif is similar to the one prevalent in the southwestern part of the United States, even though the walls are made of cordwood.

Elevation and cutaway.

Floor plan

The vigas or pole rafters are twelve inches in diameter and span twelve feet with the ends exposed in the traditional manner. All rooms face the courtyard, which contains plantings and an open sitting area. A stairway takes you to the roof level from the courtyard, so it is one of the forms of egress. The diameter of the exterior wall is forty-four feet. The rooms are all twelve feet deep, and the courtyard is twenty feet in diameter. The rooms that are living areas, such as the living/dining area and kitchen, are all glazed and face the courtyard. Bedrooms, the bathroom and the utility rooms have a privacy corridor at the front and smaller windows at the six-foot level to give privacy, while still providing light for the rooms. The interior supports for the roof are posts and girders.

The courtyard design presented here would work best in warm climates or in areas that do not receive heavy snowfall. The courtyard would tend to fill up without some protection in a heavy snowfall area. The same soil-pipe system installed in the Starlight Earth Lodge would work in this design. Recycled

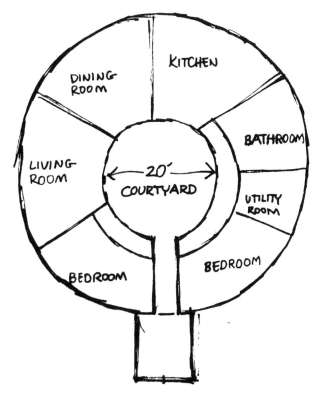

Possible floor plan for the Cordwood Courtyard.

multiple-paned storm windows would work well for the glazing that faces the courtyard. This would also allow the tops of the windows to open, forming thermal chimneys to draw the cool air from the perimeter wall.

Cool air ventilation system.

Excavation/wall construction

The excavation should be done the same way as for the Starlight Earth Lodge. Be sure that this excavation is large enough to allow work space between the outside of the perimeter wall and the earth. This structure is a mixture of post-and-beam construction and standard footings. The outer perimeter wall of cordwood rests on a standard footing that should be engineered to withstand the load to be imposed by the structure. The huge perimeter wall of this circle would normally be a major portion of the shell's expense when using cement or block, but the use of cordwood greatly reduces the cost of its construction and is much cheaper than renting forms and buying ready-mixed cement.

Cordwood Courtyard under construction near Niobrara, Nebraska. The cordwood forms the outer wall, which is round, and the butting wall is made of rammed earth, dividing the bedroom from the utility room.

The rammed earth of the Cordwood Courtyard serves as a heat and cooling sink for the structure.

A cord of wood which measures 4 × 4 × 8 will build a section of wall that is eight feet tall, two feet thick, and eight feet long. Actually, it builds a wall slightly longer than eight feet; since I have not calculated the mortar joints, I would, however, use these figures when ordering or cutting wood since any excess wood can be used as firewood. You need approximately twenty cords of two-foot-long pieces of firewood to build the perimeter wall and a couple of divider walls.

A close-up view of a cordwood wall.

NOTE: A rammed-earth wall that utilizes six-foot-long slip forms with a wall thickness of two feet could be used here. The earth mix would be 70 percent sand, 30 percent clay dirt and one shovel of portland cement for every 17 shovels of the sand–soil mix. This material, when rammed into the form until it will no longer compact, is just as strong in a circle as is the cordwood. The exterior would have to be insulated as well as waterproofed, using the rammed-earth method.

Rammed earth is easily made, using the above proportions of cement and sand.

This would allow this mass of a wall to act as a heating and cooling sink. A rammed-earth wall would store over four million BTU's of heat, thus virtually making the structure independent of a back-up heating system. The cordwood does not need the insulation because wood is self-insulating, but an additional mass has to be built into the cordwood structure to store heat since the wood does not act as a heat sink.

Constructing the roof

The interior support system for the roof is essentially a post-and-girder system. Twelve-inch-diameter posts support twelve-inch-diameter girders that are eight feet in length. The girders in turn support the roof beams, which in turn support the roof rafters and planking. Pine, oak, buckeye or willow will all serve well for these supports. Tailings or roughly cut one-inch-thick lumber can be used for the planking (adding warmth and authenticity) and may be available at a local sawmill. Roof insulation is unnecessary since the wood is self-insulating.

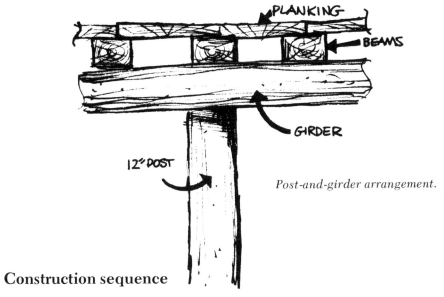

Post-and-girder arrangement.

Construction sequence

- ◆ excavation
- ◆ perimeter wall-footing
- ◆ perimeter wall (cordwood or rammed earth)
- ◆ front support system and roof (posts, girders, beams, rafters and planking)
- ◆ waterproofing and backfill (polyethylene sheets—two six-mil, then two inches of gravel for drainfield with drain tile and soil-pipe at bottom of wall)
- ◆ six to eight inches of sod on roof
- ◆ glazing of inside courtyard wall
- ◆ interior finish
- ◆ landscaping (including courtyard)

NOTE: Cordwood or rammed earth can only be used underground when the design is round or curved, since neither of these materials is self-tensioning and must be compressed for strength. If rammed earth is used on this outer wall, I would have it sprayed on the outside with two inches of polyurethane foam and then waterproofed with two six-mil sheets of polyethylene. This would not only keep the rammed earth dry, but it will isolate the wall from soil contact, thus creating a huge thermal storage unit. The cordwood outer wall will not need insulating, only waterproofing. Internal mass for thermal storage can be created in the cordwood system by using two-foot-thick, rammed-earth dividing walls inside the structure. These walls will store cool air as well as heat, and are efficient accoustical barriers. Cordwood or rammed-earth dividing walls will maintain continuity of design and appearance.

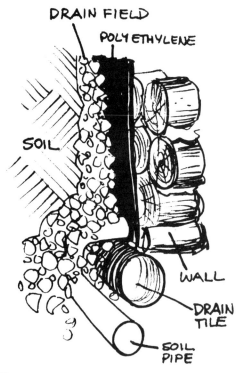

Drainage, waterproofing and soil pipe arrangement.

Garden area

The courtyard garden provides a separate world of its own and has a large enough diameter to enable the occupants to see beyond the perimeter of their dwelling to the outside world. Rammed-earth sculptures in the courtyard that could be utilized for sitting and cooking in the summer would make this area functional and visually interesting. Benches, tables, elevated areas and a rammed-earth fireplace would make the courtyard an extension of the dwelling.

Possible fireplace and bench placement.

15
Easy Arch System

Since the roof system for most underground houses represents a major portion of the cost of the structural shell, the arch presents one of the most economic methods of combining the walls and roof into an integrated clear-span system. The arch is one of the strongest systems for enclosing an area, and works best under compression, gaining strength from an evenly distributed load. Materials for creating an arched system are many and varied. The plan to be discussed here incorporates three arches that are integrated into a modest-sized dwelling for a couple or small family.

Materials

The 760 square feet of floor space in this design are spanned by arches created out of recycled rebar, metal lath and wire, and covered by a thin shell of concrete. Many construction firms sell their scrap metal to salvage yards that in turn sell it by the pound. Reinforcing bars and metal lath that have not been used on a job are often sold in this manner. If you are willing to invest your time, you can occasionally pick up materials out of trash dumpers located on building sites at no cost.

Forming the arches

One-half-inch-diameter rebar, in ten-foot lengths, is planted in the concrete footings. After the footing cement is set, the bars are bent to form arches on

One type of facade for the Easy Arch home.

one-foot centers. These arches are wired together at the center and are covered with metal lath. The lath is then covered with a stiff mix of cement. The cement is poured on the top of the rod-and-lath form and pulled downward in a two-inch thickness. If you are good at shopping around, you might be able to keep the shell cost about equal to the price of the thin shell of concrete.

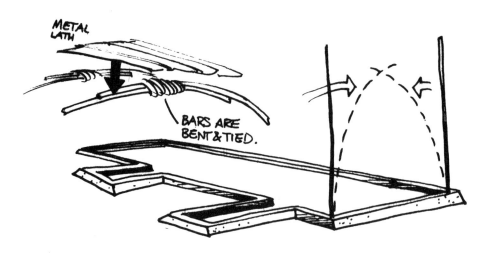

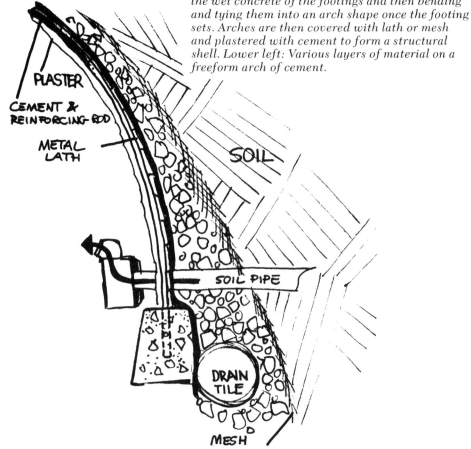

Above: Any freeform concrete arch can be fabricated by inserting reinforcing bars in the wet concrete of the footings and then bending and tying them into an arch shape once the footing sets. Arches are then covered with lath or mesh and plastered with cement to form a structural shell. Lower left: Various layers of material on a freeform arch of cement.

The interior

This design and its floor plan are very simple to construct and allow for plenty of light and functional space. Consider the following when planning the construction of your arch home:

- An interior that is divided with fieldstone walls will store heat and cold, as well as look nice and be inexpensive.
- The use of a rammed-earth floor is another way to reduce costs without sacrificing appearance.
- The interior walls can be plastered with dry-wall compound, and the wall surface can be smooth or textured.
- Wiring and plumbing can be attached to the skeleton prior to surfacing. This method allows visual inspection of all details before covering.

Heating and cooling systems

Glass at the front of each of the three arches can be comprised of recycled wood-framed storms, windows, door or store glass. The glass should provide adequate passive solar gain to heat the interior. Interior mass, such as a Russian fireplace or stone dividing walls, will help store the solar gain in the winter for use at night or on cloudy winter days. Cooling should be accomplished with the use of soil-pipe and thermal chimneys as indicated on the floor plan. This soil pipe can be combined with the drain field and enter the structure along the north wall. This cool air will be drawn towards the glazed openings (by the escaping hot air) through windows at these points.

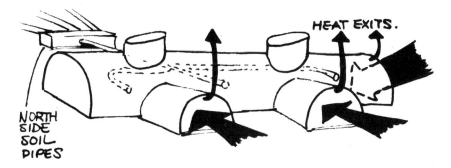

Cooling, dehumidification and ventilation system for arched structures.

Construction sequence

- excavation
- footings. Rebar must be placed while cement is still wet.
- Bend rebar arches and wire, and fasten metal lath to arches. (Be sure all shell openings are included at this point: light well, sewer vent, flue, etc.)
- Fasten electrical conduit and plumbing to skeleton.
- Cover skeleton with two inches of concrete.
- Insulate with sprayed-on polyurethane foam (two or three inches at top and graded to one inch or less at the bottom).
- Waterproof with double sheets of polyethylene six-mil thick.

- ◆ Place soil pipe (twelve inches in diameter, two pipes that are one hundred feet in length each). These should be above the footing drains and enter the rear wall through another twelve-inch pipe plenum.
- ◆ Backfill with a two-inch gravel drain field over entire structure. (The dirt goes over the drain field but should be separated from the gravel with some screening material until backfill settles; this keeps the gravel from being plugged by dirt.)
- ◆ Complete interior finish of plaster with field-stone divider walls and rammed-earth floor.
- ◆ Glaze with recycled wooden storm windows, doors, etc.
- ◆ Landscape.

16
Double Hex

The Double Hex uses cement in the most nonskilled way possible. Ten-by-ten-foot wall sections are cast on the ground, then tipped up onto footings that have an inside lip that acts as a retainer and water-sealer.

Possible facade of a hexagon tip-up wall structure.

Floor plan

As the floor plan indicates, the double module is a very simple and straightforward design that can be constructed by one person. The double units can be joined into four by simply butting them back to back. The additional units are all interreinforcing and need no additional supporting systems. The roof systems are all monolithic pours with joist beams. The forms for the roof are 4 by 8 plywood sheets with troughs formed out of 2 by 10s. This form plan is supported by a simple truss arrangement that is made up of 2 by 4s that are braced on four-foot centers. Nothing is nailed, except the truss troughs; therefore, any material that is not going to be used in the structure after the forms are torn apart can be sold as new material (same system is used as that of the Convertible Crescent).

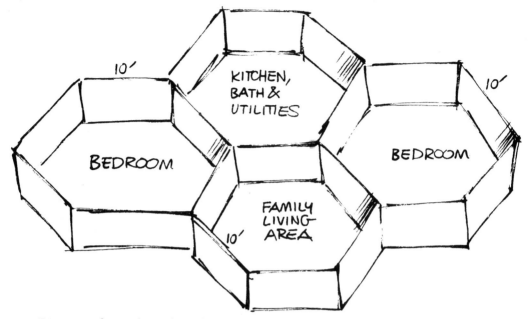

Rooms can be easily made and added onto the hexagon tip-up wall construction.

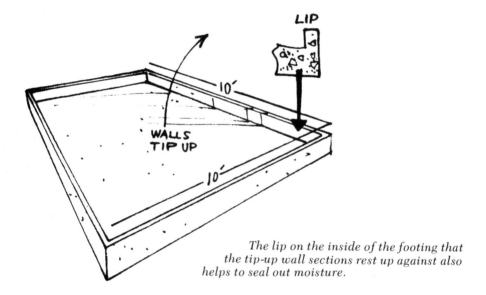

The lip on the inside of the footing that the tip-up wall sections rest up against also helps to seal out moisture.

Tip-up wall sections

All windows, doors and other wall openings can be formed into the tip-up wall section. These tip-up wall units use gravity instead of vertical forms, thereby allowing units to be done individually instead of pouring the entire wall system at one time. While the concrete is wet, the surfaces of each unit may be textured, rocked or treated with wood or any other surface treatment. Placing colorful rock or other decorative material into the form is easy to do since the units are formed by using 2 by 10s in a ten-by-ten square frame on the ground.

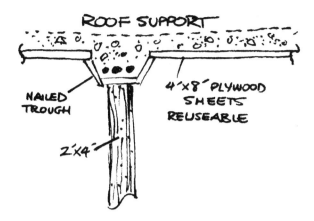

A poured-in-place, monolithic joist beam and deck are formed by using ¾-inch-thick 4 by 8 sheets of plywood with polyethylene cover, supported by a "kick-out" 2 by 4 truss system.

Tip-up wall sections are formed on the ground so that doors and windows can be formed by placing their frames inside the main form. Concrete is then poured around the inside forms.

The inside portion of a tip-up wall section can be textured by placing the textured material on the face of the wet concrete and allowing it to set.

A trough is formed in the top of each unit, using a piece of four-inch-diameter plastic pipe that is cut in half and nailed to the top of each form inside. When poured, this leaves a four-inch rounded trough. When the units are tipped up into place, a five-foot-long one-inch rebar is placed into half of the trough of one unit and then bent to make the corner and run to the center of the adjacent unit. The troughs are poured full when all of the units are in place. A bond beam is formed around each module, locking the modules together at the top. This makes a very strong, integrated six-sided module. Be sure to place all conduit and other service elements in the units before pouring them.

Illustration shows how a channel can be formed at the top and sides of a tip-up wall section. These channels are then filled with reinforcing rods and cement to lock the units together.

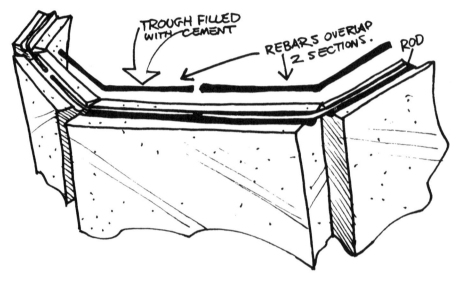

Method of connecting the tip-up wall sections in hexagon or other multi-sided structures is shown above. The top reinforcing rods are placed in the channel at the top of each unit and then covered with cement.

Construction sequence

- ◆ excavation
- ◆ footings
- ◆ drain tile
- ◆ soil pipe and plenum pipe
- ◆ tip-up wall units
- ◆ bond beam
- ◆ Place forms for the roof, pour the roof and remove the forms.
- ◆ Waterproof and insulate.
- ◆ Backfill with French drain field around entire structure.
- ◆ Glaze.
- ◆ Finish the interior.
- ◆ Landscape.

I would insulate this building by spraying on polyurethane foam, which seals all of the cracks, and covering the structure with double layers of polyethylene six-mil sheets to waterproof over the foam.

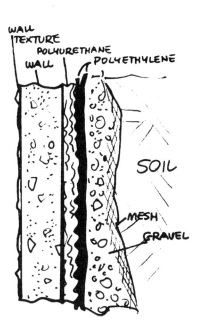

Layers of material are applied in sequence once the shell of a structure is fabricated from sprayed-on polyurethane that is placed over wire mesh.

NOTE: Always insulate according to the climate. In a very cold climate, I would insulate the entire structure, including under the floor and footings. A hot or warmer climate may only need insulation on the roof and down the walls to the point where the ground becomes temperate.

17
Wondrous Warren

The Wondrous Warren is an arched type design that utilizes a large culvert structure produced by steel fabrication firms. The snug appearance of this design allows it to be placed into almost any terrain. It will fit into mountainous and hilly or flat land, and it can be faced in almost any direction without too much of a thermal penalty. The round light shafts can always be faced south to pick up heat for winter. These light shafts have collectors in the back that transfer the warm air to rock storage under the floor. Gravity forces this rock to heat each unit from the bottom through floor registers.

Shell structure

The shell of each unit is fabricated on site. By building a rectangular box inside the shell, the interior is left uncluttered. The space between the shell and the box can be used for closets, chases for utilities, rock storage for heat, and a crawl space above for electrical wiring, etc. Inserts can be made by using 2 by 4s and paneling or plywood, which can be finished on the inside with any texture you choose. Since they do not have to be insulated or sheeted over on the exterior, the inserts can be built at a low cost.

Use of steel culverts

The steel culvert tube is insulated on the exterior by spraying it with a couple of inches of polyurethane foam, which is then wrapped in two sheets of six-mil polyethylene.

No footings are required for this structure since the structure just floats on a bed of gravel that serves as its drain field. This is an excellent structure for placing into the ground on a building site where the soil is expansive or otherwise unstable. The weight or load distribution is spread over the entire bottom of each unit. Although these steel tubes are not inexpensive, their cost compares favorably with concrete or other materials.

Once the tubes are sealed with polyurethane foam and plastic sheeting, the structure becomes a permanently sealed and livable home. The tubes are oval in shape, and can be ordered in almost any length you need. Remember that if you get a wide one, it will also be tall. Two tubes, 14 feet wide and 12 feet tall, make a nice home when placed side by side with a connecting tube for a hallway. A mobile home manufacturer could easily fabricate the habitat inserts in the factory, tow them to the site like a mobile home, and insert them directly into the tube.

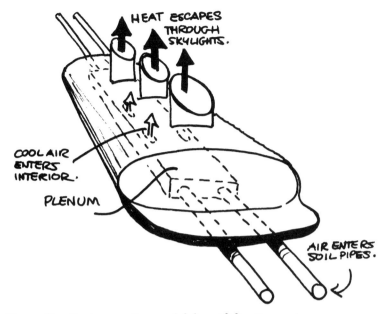

Above: Ventilation, cooling and dehumidification system.
Below: Wondrous Warrens, Freedom Tunnels and similar structures can be constructed and used as schools, hospitals, motels, offices, etc.

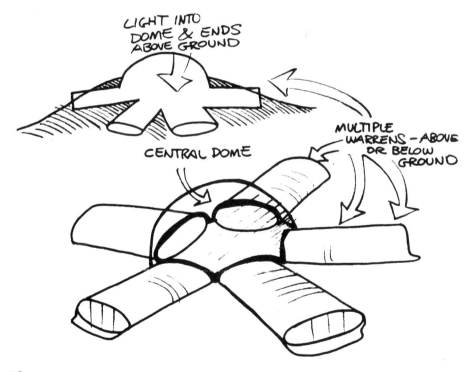

Plenum

A plenum should run the entire length of the floor of the insert. This plenum would then be fed by two soil pipes at either end. The light wells would serve as thermal chimneys for draw. The skylights or wells serve to heat, light and cool.

Three separate Wondrous Warrens can be built or one dwelling can have connecting passageways to several levels.

Construction sequence:

- ◆ excavation
- ◆ gravel bed. (Be sure all plumbing, wiring and other utilities are in at this point.)
- ◆ Order in the steel shell and have it fabricated in the excavation. Have the bottom plates sprayed with polyurethane foam insulation and lay down a double layer of polyethylene sheets for the shell to rest on. These sheets will be drawn up the sides and overlapped with the top sheets.
- ◆ Build the living insert over a bed of rock at the bottom of the shell. Be sure to fasten wiring conduit, plumbing and other utilities to the inserts outside.
- ◆ After insulating and waterproofing, backfill with a surrounding gravel field and dirt.
- ◆ Landscape.

Multiple Wondrous Warren units could be arranged around a hub which could be covered with a geodesic dome. These units would make a great elementary school that would be almost entirely energy independent. The units could be placed on flat ground and be bermed over, or they could be put into a hilltop. You can use a small unit for a cabin or studio, or you can have a rambling six-thousand-square-foot home.

18
The Stairway House

This house uses a long, gently sloping hill to its best advantage. It provides multilevel living with no barriers, especially welcome for the handicapped. With this design, you can have the construction subcontracted. Any basement contractor could pour this structure on its double-beam footing.

Costs

The design of this 1,800-square-foot house with its well-lit levels, can be built for three quarters of the price of a conventional frame home of equal size. Using the equity one would have in a present home, the Stairway House could probably be built and paid for. The long-term energy savings in a home like this would also pay for it when compared to the energy costs for a conventional structure above ground of equal size.

Double-beam footing

The roof sections are 15-foot-long Flexicore units. The supports for these roof units are two poured beams with supporting columns and the back supporting wall. The double-beam footing system is one of the unique features of this design which not only makes the structure strong, but anchors it to the hillside as well. This is another design that works well in expansive soil. The double-beam footing holds each living level in place and keeps the earth from shoving the upper or back units into the bottom one. Although this structure may appeal more to the people interested in a more conventional or modern design, it still can be built for less than a comparable frame home.

Heating and cooling systems

The glass at the front of each level allows light and heat at each stage, while giving an unparalleled view of the surrounding landscape. This glass serves as a mini-greenhouse for each level, allowing abundant plant and gardening possibilities. A back-up heating system, such as a woodburning stove or Russian fireplace, would heat all the upper levels by simple gravity air flow. There should be enough structural mass in this design to make a back-up system unnecessary, but many people still like the idea of being able to have a fire in the winter for its cheer and comforting effect.

Cooling and dehumidification can be achieved with equal ease in the summer. Four one-hundred-foot lengths of soil pipe enter the structure at four points on a common plenum that runs between the double support beams under all three levels. Hot air is vented from the windows of each level in proportion to inlet capabilities for that level.

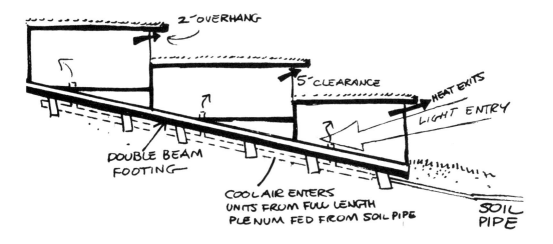

Cutaway of the Stairway House illustrates its multilevel quality.

Construction sequence

- ◆ Special excavation for double-beam footing and lower portion of walls. This excavation slopes with the hill, pouring of beams and floors as monolithic unit. Be sure to leave exposed tie-bar for the wall which will be poured later, and insulate footings and floor if necessary for your climate.
- ◆ Pour the walls and posts and beams that support the Flexicore roof panels. Be sure all conduit and other utilities are in the wall forms before pouring.
- ◆ Place the roof units.
- ◆ Insulate and waterproof. I would recommend spraying on polyurethane foam insulation and using double six-mil sheets of polyethylene as a waterproofer.
- ◆ Place drain tile and gravel drain field, and begin backfill.
- ◆ Glaze and close in the structure.
- ◆ Complete interior finish. Be sure soil pipes and plenum are in place under and between the double support beams of the floor. Place one foot of sod on top of each of the units.
- ◆ Landscape.

19
Geotecture: Self-Supporting Tunnelled Space

If care is not taken in the design of a tunnelled space and its habitat systems, dankness, darkness and gloominess can result. I have tunnelled personal camping shelters and weekend hideouts, all of which were dry, warm and filled with light. A tunnelled dwelling is quiet, safe and secure.

Site selection

Although some tunnelled shelters have probably caved in at some time, there is no reason one should if care has been taken to locate a good digging site and some basic tunnelling rules have been followed. Contrary to most beliefs, you do not have to tunnel into rock in order to have a self-supporting space. Any good clay soil that is uniform in makeup and free of large root systems, rock or underground water is satisfactory. The ideal site would be one with a vertical face to tunnel into and about a 70 percent sand and 30 percent clay/soil composition. The face has to have enough height for at least six feet of overburden above the highest point of the arched portal. A mattock, tile spade, cleaning shovel and wheelbarrow are the only construction tools that are necessary.

Arch construction

Establishing the arch is the first priority when tunnelling any space. The arch is what allows the overburden to be held in place and the desired space to be spanned safely. I usually make a template, using one-half-inch-diameter rebar. Bend the rebar in the shape of the desired opening, and wire a piece across the bottom to maintain its shape. Use this template to establish the space and keep it uniform. It is the tremendous compression of the arch by the weight of the overburden that makes the vault self-supporting and strong. The more overburden, the better. Decide the width of the portal and the height of the side walls; the arc of the arch will then have to be determined. Generally speaking, the greater the width to be spanned, the greater the arc from where it begins at the top of the side walls.

A well-formulated soil that is uniform and under good compression will span just about any space if the arc is correct. An eight-foot span should have at least a two-foot arch; a ten-foot span, a three-foot arch; a twelve-foot span, a minimum of a five-foot arch. The arch to be formed should start at the shoulder of the wall and make a uniform curve to the highest point of its arc. If there is any question about the soil or the feasibility of tunnelling, contact a qualified geologist or mining engineer.

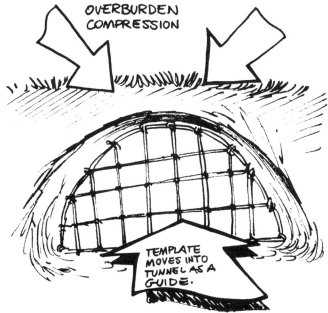

Use of the template in maintaining the shape of a tunnelled space is illustrated above.

Soil samples

Sample augers of soil taken at the depth the tunnel will be dug, and along its path, should be taken. These samples taken at four-foot intervals will give you an indication of whether or not you are going to run into such obstacles as loose sand, rock or water. If the route to be tunnelled is too cluttered, find another area.

Refrigeration, electricity

A tunnelled root cellar is an excellent addition to any earth-sheltered home and will pay for itself in produce storage many times over. One that is tunnelled directly off a kitchen would create a convenient storage area for many items that often are refrigerated. A root cellar could almost do away with the need for refrigeration in all but a few instances, thus making one of the small 12-volt refrigerated camping units about all of the refrigeration needed. A 500-to-1000 watt generator or a wind generator would then be practical for all of the basic electrical needs of an earth-sheltered home.

Heating and cooling systems

Be sure to install soil-pipe ventilation and a solar chimney. One hundred feet of a twelve-inch-diameter soil pipe entering the space at the front, in or near the floor, will adequately dehumidify and cool an area that is eight by sixteen feet. The soil pipe has to be coupled with a solar chimney that exits at the rear and top of the interior space. The solar chimney will draw the air more rapidly as heat rises, thus keeping a fairly constant temperature inside. The soil-pipe ventilation will keep the interior dry and odor-free.

Mother Earth staff members are shown digging the shelter. The rebar template beside the opening was used to keep the space uniform.

The interior

The interior surface of any tunnelled space may be plastered. Simply form rebar arches that rest on bricks on the floor and conform to the walls and roof of the space, fasten metal lath or a fine mesh screen to these arches, and then plaster

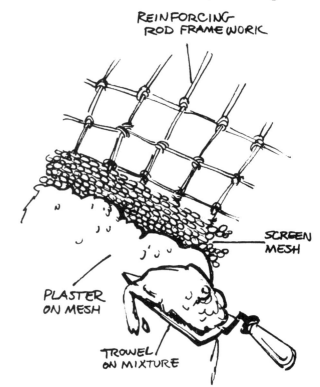

Finishing interior surfaces of a tunnelled space.

them over. The bars can be about two feet apart. The plaster mix can be one part cement to five parts milled soil. If a little sand is needed, add it as necessary. This mix retains the color of the soil and gives the interior a natural feel and appearance.

Other interior advantages

- ◆ Windows and doors can be made of recycled storm windows and doors with multiple panes, which can be inexpensively purchased from salvage dealers.
- ◆ Light wells at regular intervals, coupled with light-colored or reflective interior surfaces, will make the interior as pleasant as any surface or near surface home.
- ◆ The ability to carve out extra space whenever needed for storage or an extra room makes tunnelling an ideal form of construction.
- ◆ Benches, bed platforms, tables, fireplaces, end tables and various levels may all be sculpted out of the soil and plastered over, thus making the interior basically sculptural. You can decorate the space as you wish: One tunnelled home I've seen was furnished with antique furniture and cut-glass chandeliers, and made me feel as if I were in a Victorian living room.
- ◆ Wiring and basic plumbing will probably be about your only expense, with the exception of the doors.
- ◆ The floors may be rammed earth, adobe or recycled brick.

The main advantage to tunnelled space is that there are no structural shell costs, since, of course, there is no structural shell.

Door frames

Plank doors that conform to the arch make attractive and secure entries. Green 1 by 6s, laminated and bent to the shape of the arch, make good door mountings. This frame can be secured to the earth sides by placing the frame about two or three feet back into the tunnel and then fastening it in place with long toggle bolts.

Soil erosion

The face of the bank that you tunnelled into can also be plastered at the various portals and doors in order to retard soil erosion. These surfaces are plastered in similar fashion to the interior surfaces, except the metal lath or fine mesh wire is pinned directly to the face by the use of long staples formed out of number nine wire. This plastering should only be around the openings for the tunnelled shelter and not over the entire face of the embankment. A slurry of grass seed should be sprayed on the bank so that the grass roots will hold the soil against any water or wind erosion.

One advantage to tunnelled space is its low visibility: Once the face has been penetrated a few feet, all construction activity is concealed. The soil that is removed can be distributed as fill about the area and will blend into the surrounding topography. Tunnelled space can meet codes, but some areas have local rulings that would hinder the use of this type of space as a place of permanent habitation.

20
Energy-Efficient Systems

There are alternative methods to using conventional energy sources (gas and electricity) in your underground home that are efficient and less expensive in the long run (some of which can be used in surface-structure houses). These include using a wind generator to produce electricity, installing soil pipe and solar chimneys to eliminate the need for traditional air-conditioning, and using a composting toilet to eliminate the waste of water associated with conventional toilets.

Wind-generating system

The simpler and smaller you can keep your wind-generating system, the better. My friend, Rob Roy, who built an underground house in northern New York State, powers that house and a cabin with a 500-watt Sensenbaugh generator. That is a relatively small machine, but it handily meets his total electrical needs. He has sized his load and leveled it by not creating large peak loads.

Becoming your own power company is not all that difficult. The ideal system would be to start with one 1500-watt machine on a tower that clears any wind obstructions. The machine should be located as near to the point of use as possible to prevent line losses. Direct current (DC) tends to lose a lot of power if transmitted any distance (which is why in the early days of electrical power, when each town had its own DC generating plant, the houses nearest the plant were lit the brightest, while those on the outskirts looked dim in comparison).

Install a battery bank of deep-cycle garden tractor batteries, and wire the unit to supply the electric needs of one or two necessary household functions (perhaps the furnace fan and the lights in the bedrooms and other areas that you consider secondary).

I suggest using one wind generator at first and then getting a second one. Getting used to utilizing a wind generator allows you to gain confidence in having that type of alternative energy source, while being cost-effective. Two smaller machines are better than one large one because the large one will not reach its rated output in low-velocity winds. The cost of one large seven-kilowatt machine is about twice the cost of two smaller units. Should one machine need repair or maintenance work, the other one could still supply the basic needs of your battery banks. The smaller units reach their peak output in much lower velocity winds; therefore, they make more use of the lower speed, but more frequent winds. The large machine will not even generate usable electricity at the lower wind speeds that the smaller machines utilize. The reason I suggested you run your furnace fan as an experiment with your first machine is that when the power goes out during a storm, that is the time you need the heat distribution the most. High winds during a storm will create the most electricity through your wind generator. Wiring additional lights in rooms that are used most will keep you from having to burn candles during these winter power outages.

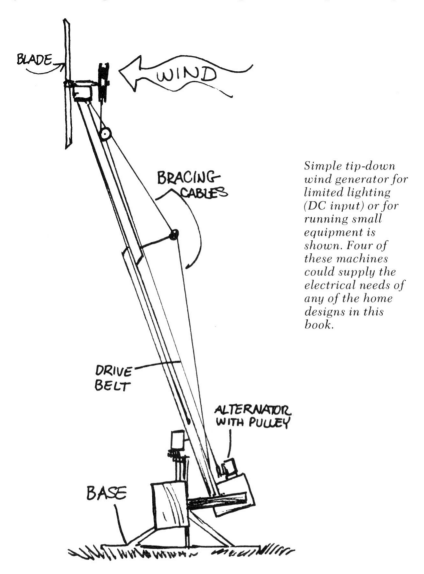

Simple tip-down wind generator for limited lighting (DC input) or for running small equipment is shown. Four of these machines could supply the electrical needs of any of the home designs in this book.

The steps to creating a successful wind-generating system are as follows:

1. Reduce your peak electric load to at least 3,000 watts. Do this by using alternative methods for many of the unnecessary uses of electricity. (Hang clothes outside, wash dishes by hand, cook with gas, heat water with gas and solar energy, use a carpet sweeper and a broom, hand-mix most recipes, install soil pipes and thermal chimneys to avoid having to use air conditioning, use sharp knives instead of electric ones, and use a root cellar and small portable 12-volt camping refrigerator or gas refrigerator.)

2. Level your electric load by doing high-wattage jobs when other appliances are not in demand.
3. Start with one rebuilt 1500-watt machine, get used to using it, and put the second one into operation.
4. Use DC, 12-volt current (deep-cycle garden tractor batteries), and install the towers as near to the house as possible.
5. Buy 12-volt appliances that are efficient. (Twelve-volt DC lights last a lot longer than alternating current [AC] bulbs and have more lumens per watt.) Inverters are available for televisions, radios and other appliances. Many appliances that you presently own can be wired easily to work on 12-volt DC. Go to any small-appliance repair shop and tell them that you want to operate your appliances on 12-volt DC.

Gas and hot water

Gas and water are additional resources to be considered when building an underground house. Building underground and using any of the energy-efficient designs in this book will automatically reduce or eliminate any need for gas as a heating source. This leaves only a couple of essential uses for either natural or bottled gas.

The main use of gas will be for cooking, water-heating and, possibly, for refrigeration. The gas for water-heating can be greatly reduced by using solar energy as the primary source for heating water. There is little need of very hot water, anyway, so solar energy is ideal. About twenty percent of a household's gas bill usually goes for heating water, much of which is unnecessary. Dishwashing is about the only job that requires really hot water. The fabrics in clothes today wash just as well in cold or warm water as they do in hot water.

If you are going to heat water with gas, there are several source-type water heaters available that are far superior to large conventional hot-water heaters. A

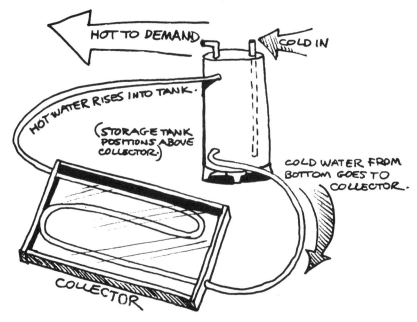

A thermal-siphon hot-water heater that is solar-powered can be easily built.

tremendous waste of gas goes into heating fifty gallons of water and keeping it hot until it is used. It makes far more sense to use a source heater on the line that will supply the water as needed, and as hot or cool as needed. These devices have a thermostat and regulator that allow a precise temperature setting and a regulated flow at that temperature. The gas is only used when there is a hot-water demand, thereby saving many cubic feet of gas. A fine mist-type head on the shower decreases the amount of water used. That, coupled with the source-type heater, will save water and gas. A simple thermal syphon solar hot-water heater is illustrated and will work well in your underground greenhouse or solarium.

Composting toilet

Water is taken for granted by most of us, but it should not be. Fresh water supplies are being threatened more and more by pollution and wasteful usage. I have recommended the use of composting toilets in lieu of flush toilets and septic or sewage treatment to save water. A composting toilet is easy to make, or it can be purchased. It does not produce an unpleasant odor, since all odor is vented up a stack exactly like the sewer stack in a conventional system. The solids are decomposed rapidly through a pasteurizing chamber that is kept at a constant temperature of about ninety degrees Fahrenheit. You can run this heating element and the low-watt vent fan from a single battery and a solar voltaic unit mounted on the roof. If you prefer, you can have them run off a wind generator and battery bank since the heating element and fan require very little energy.

If you are building your underground home in the country, a composting toilet eliminates the need for a septic system and a lateral system (which is a considerable expense). The grey water from other household uses such as laundry and bathing can be run off into a charcoal-and-gravel leach field and used on the garden or lawn.

The construction of a composting toilet is not difficult. There are two types of these biological recycling units. One type is a two-piece unit. The top unit is the equivalent of the toilet; the second part is a large pasteurizing chamber that is usually located on a separate lower level. In surface homes, the pasteurization chamber would be located in the basement, while it would probably have to be located in an excavation or prepared chamber below the toilet chamber in an underground house. This type of recycling unit allows large quantities of household biodegradable waste to be treated. Because of the larger bulk of degradable material, the temperature (over ninety degrees Fahrenheit) is maintained by the mass itself without an auxiliary heating unit. It is still advisable to include a low-watt vent fan in the stack to evacuate any resulting odor.

The second type of unit is the portable recycling chamber. This type of composting toilet is a one piece unit that requires a heating coil to maintain a temperature in excess of ninety degrees Fahrenheit in order to sustain the biological action. Oxygen and heat rapidly reduce the human and kitchen waste into a small amount of nutrient-rich humus that can be used in the garden.

The smaller, portable unit requires no installation other than a hole in the roof for the vent stack. Considering the cost of a septic system or a plumbing hookup to the sewer line, the cost of a composting toilet is reasonable. The tremendous amount of water saved is to be calculated, especially in low-water areas and in a city or town with a municipal water supply that charges its customers.

If you want to build one of these two types of units, the actual cost would be considerably less than the cost for a comparable commercial one. The main construction materials would be lumber and a simple nonbiodegradable liner of plastic sheeting that would protect the wood. You must, however, be sure that oxygen and ventilation are constantly supplied to the system and the temperature of over ninety degrees Fahrenheit is maintained. PVC pipe will serve as venting and connecting tubes. The use of scrap lumber, twenty-mil polyethylene and pipe with a low-watt vent fan should keep the cost of this unit low.

It may take some time and patience to educate your family to accept this mode of waste recycling. Let me assure you that it is perfectly sanitary and has been accepted by a number of small communities as their sole means of waste management. Municipal and larger health departments might not approve of it, but this system is being regarded as the best answer to sewage disposal and water wasting by a rapidly growing segment of the population, especially in water-restricted areas.

One last thought regarding water: Many areas have water where you may want to build, but the water is either polluted or contains salt. If the supply is not too polluted for bathing and washing clothes, then the best way to purify your actual drinking and cooking water is to distill it. Distillation evaporates and recondenses the water, removing bacteria, salt and other impurities. Simply boiling the water will kill bacteria and viruses, but will not remove salt, minerals and other deposits.

21
Landscaping

Landscaping should be considered and planned before starting with the excavation. Not only should slope and sun orientation be considered, but what method of excavating will do the least damage to the surrounding ecostructure should also be considered. If you are excavating in a rural setting, the land already has a natural balance unless it has been ruined by man. Insects and their predators, gophers, snakes, rabbits, coyotes, rats, mice and owls all have struck a balance that will have to be maintained. Having lived on farms and in rural areas a good portion of my life, I have learned to have a lot of respect for this balance.

The last place I lived had an excellent balance of birds, insects, snakes and other forms of life. My neighbor cut and burned almost all the trees that grew on his field that bordered my property. I immediately had an increase in insect population because all the birds that inhabited the trees disappeared. Instead of using chemical sprays to combat the large number of insects, I bought two hundred young chickens and turned them out in the fields to reduce the increased insect population. Consequently, little damage was done to the ecostructure of the building site, and I had chicken to eat all winter.

If the area in which you are building (especially the actual building site) is almost barren and uninteresting in visual appearance, there are a number of things that can be done economically that will improve the appearance and still blend with the natural surroundings.

Uses of landscaping

When landscaping, there are several things you need to accomplish other than just adding beauty to your home. Privacy can be enhanced, security can be increased, special use areas can be created, and thermal efficiency can be increased—all with plants and earth. Deciduous trees should be planted on the sun side of the dwelling. They will provide an air-conditioning effect while allowing the sun to heat the dwelling in the winter. If your home is going to face southward with a greenhouse exposure, the land will need some immediate protection during the first summer to break hot winds and direct radiation. If there are no trees large enough to provide this protection, a good way to provide it quickly with smaller trees is to plant them on a berm. This raises their shade level immediately, while giving the front of the structure a lower wind profile.

Enhancing privacy

Privacy and special use areas may be created around the dwelling through the use of additional berms and mounds with hedges and shrubs. I have seen homes with concealed patios that were created with the use of berms and hedges. Such areas can be arranged to further shield the structure from windchill in winter.

Increasing security

Security in a rural setting may be more difficult to create than in an urban area, simply due to the isolated location. Concealment of the driveway and the actual shelter from road view is desirable. Just as this may seem to make you more vulnerable because no on else could see an intruder's activities, it is also a deterrent to this same activity. Limiting access to just the driveway is easily done by the use of plants. Trifoliate Orange, a very dense and thorny bush, grows to twelve or thirteen feet and is totally impenetrable. These bushes, when planted as a perimeter barrier, will assure that no one will randomly hike across your land. These same bushes can be planted under windows to assure that no one will enter the structure by the window route.

A rural setting should be kept in tune with the surrounding topography as much as possible. The use of wild or native grasses is best. Plant a native grass that is hardy for your particular climate. Let it grow and do not worry about mowing it. Plan walkways and sitting areas that are self-maintaining.

A landscape for an urban home requires a little more planning, but it can still be a natural and self-maintaining entity. I would use a series of terraces, berms and mounds to create privacy areas. Rather than plant a yard, I would plant trees and shrubs. Flowers and vegetables can also fill in the landscape. Properly planned, your habitat will be the envy of neighbors, creating a unique, personal and private living area.

Thermal efficiency

Your dwelling can be made more thermally efficient as a result of your landscaping procedures. Use needle trees to create winter windbreaks, deciduous trees for summer shade, and bushes to reroute prevailing winds in both summer and winter. Morning glories grow rapidly on trellises and will shade over patios and other privacy areas. In the winter, they will die and allow the structure to gain radiant heat from the sun.

The thicker the grass and its root system on the roof of your home, the more thermally efficient it will be. Grass is one of the best cellulose insulators available. A thick grass on the roof in summer will transpirate and cool, while the dormant roots and dead tops form air spaces that keep freezing to a minimum in the winter. Talk to the County Agent, Soil Conservation Service or a good local nurseryman. Many times a local college or university will have an agronomist or soil expert on staff that can give you useful information on plantings.

Gardening

A building site in the country offers an excellent opportunity to grow a large portion of your own produce. This not only saves money, but provides your family with a higher quality of food. Planting your own garden and balancing the plants to protect each other from pests will sidestep most of the costs associated with gardening.

French Intensive Method

The French Intensive method of gardening not only produces from four to six times the produce per space utilized, but it is also a very eye-appealing part of the landscaping. Basically, the method involves the creation of growing beds that are double-dug to a depth of about twenty-four inches and then mounded up. These are permanent beds about five feet in width and twenty feet in length. The arrangement of the mounded beds allows permanent walkways between them for ease of harvesting. The use of systematic mulching and turning the beds twice a year keeps the soil extremely loose and allows the plant roots to deeply penetrate the soil with no resistance. The highly enriched soil stays moist through the winter with the heavy mulch cover and is seldom frozen in the spring, thereby allowing the seeds early germination.

The fact that the plants do not have to compete with each other for water or food allows them to be planted much closer together and mature more rapidly. This is the reason that four or six times the amount of produce can be raised on a small plot as compared to conventional gardening methods. The seeds and plants are set in the ground in precise geometric patterns that allow each individual plant to protect its neighbor from pests and weeds. The plantings are so close together that they shade out weeds.

The original bed preparation involves strenuous work, but once the beds are established, it takes only about an hour to turn each one in the fall and again in the spring. The only tools required are a spade, shovel, spading fork and a wheelbarrow. Once the beds are permanently established, there is less physical energy exerted in working them than there is in wrestling a rototiller through conventional garden soil. I personally enjoy using my hand tools early in the morning to turn my beds, as there is something very special about listening to the birds and other sounds of nature while quietly working the soil. The fact that this method does away with weed pulling, rototilling, excessive water, fertilizing, spraying and the usual time-consuming maintenance jobs associated with gardening makes it a truly enjoyable hobby and profitable venture.

The moisture is held in the soil by the heavy composting and overlapping leaf patterns. The only requirement for watering is the shineying of the leaves once a day (shineying means to mist the leaves until they glisten from the water). This is done with a special misting head that is attached to the garden hose, and the misting itself only takes about ten minutes a day to complete. It should be done early in the morning before the sun can rapidly evaporate the water.

You will be shocked at the simplicity of the shineying method and the large amount of produce that results. It gives me a tremendous sense of accomplishment to harvest such a large amount of produce from such a small space. Your friends and neighbors will be astonished when they see the amount of produce. A three-bed plot should easily supply a family of four with enough produce to get them through the winter. An underground greenhouse will supply table-fresh perishables such as tomatoes, cucumbers, melons and lettuce all winter.

Landscaping steps

Landscaping involves a cohesive, progressive plan. The following steps should be taken immediately when planning for the construction of your underground home.

1. Learn what life forms exist at the building site and in the surrounding area. Decide what types of excavating equipment and methods to use that will disturb the landscape the least.
2. Plan the end use of the excavated soil, and deposit it at that point during the digging process.
3. Plan windbreak berms and plantings, along with privacy and special use areas.

4. Test the soil and add nutrients when necessary to condition the soil for the plants that it will nourish.
5. Plan security plantings to limit entry to the property.
6. Use the French Intensive method of gardening and protect it from the wind. (This should also be planned in conjunction with a root cellar and a small greenhouse.)
7. Select plants that grow well in the climate in which you are building. Do not select plants just on preference or by what is popular.
8. Draw the entire plan on paper and follow it. This will save money and time and will insure that you have the most attractive, thermally efficient landscaping possible.
9. Include the plants and earthwork in your building budget; otherwise, the landscaping will not get done, and the overall efficiency of your underground home will suffer.

I cannot state strongly enough that the landscaping should be planned right along with the building design so that everything is integrated upon completion. Landscaping for the underground home is probably more important than for other structures. The underground home has little façade to show to the public, so it is important that the site be attractive as well as thermally efficient and useful.

Roof soil and erosion

Holding soil on the roof of the structure becomes a problem if erosion occurs. Sodding is a quick solution to this problem, but it can be very expensive. If the roof does not have a base of subsoil, there will not be much root structure to the sod. There also won't be any moisture-holding ability. Mix rye seeds with Kentucky Fescue grass seeds and you will get a quick ground cover. The cover will be permanent and have a heavy root system that will stop erosion from even the worst runoff. Plant winter wheat in the fall and interseed it with a good long-term ground cover. The wheat germinates quickly in the fall and provides immediate ground cover that will hold all winter and be in place for the heavy rains. As the wheat dies off, the long-term cover will be well rooted and ready to take over.

Below is a partial listing of ground covers you may want to consider.

Kentucky Fescue	Blue Grass
Crownvetch	Buffalo Grass
Moss Roses	Brome
Zoysia	Ground Ivy
Bluestem	Bermuda

Check the local nursery to see what ground cover will work best in your soil and climate. I tend to prefer the broad-bladed grasses that have heavy root systems. These grasses tend to stand up in the hot dry periods of summer without needing water. I never mow grass, and I plan use areas that are either bricked or surfaced by rammed earth.

22
Spatial Perceptions and Uses

In earth sheltering, we are compromising visual planes by using the northern exposure in cold climates to obtain maximum insulative factors by covering it with earth, thereby eliminating most northern window exposures.

Some preconceptions concerning underground houses

Comments such as, "I simply could not live without seeing the sun go down," are common when discussing below-surface dwellings with a person who has never been inside an earth-sheltered structure. In order to compensate for the removal of some of the expected visual planes, it is necessary to substitute other elements. Workable solutions include aquariums, murals, mirrors which reflect light into rooms, and indirect lighting and multitextured fields of reference (such as the heavy use of plants indoors).

An elderly lady who was on a tour of my home commented to me, "Why, this is really nice, you can just walk right in; we thought you would have to crawl down into something." Another woman commented that she felt claustrophobic in the back master bedroom of my home. When I asked why, she said it was because she knew she was underground. I asked her if she felt that way in her basement, and her reply was no. It seems, therefore, to basically be a matter of education and being able to train oneself to relate differently to the space underground.

If your spouse and family have any lingering doubts or questions about living underground, it would be advisable for them to talk with owners of earth-sheltered structures and, if possible, to take a tour of one. The vision of dark rooms covered with tons of earth are quickly dispelled with simple exposure and education. My daughters, who did not have preconceived notions regarding living underground, never felt that they really were underground, but merely in a light-filled and enjoyable new home.*

Noise levels

Probably the most common complaint from people living in earth-sheltered dwellings is the lack of noise from the exterior. This is a problem of adjustment, for we live in a world filled with high-decibel levels of noise (trains, jet aircraft, cars, and industrial noise). Because the use of earth in sheltering provides such a superior accoustical barrier, all of the usual outside stimuli are eliminated. If you sit alone in an earth-sheltered structure, all you will soon be able to hear will be

---

*Some favorable attributes regarding underground structures include the increased attention and learning levels recorded at Terraset Elementary School in Reston, Virginia. Teachers there are impressed not only with the tremendous energy savings, but also the increased student response. It is almost as if the students feel a part of the surrounding ecostructure through the building and respond to this stimulus. I feel that this same response happens in an earth-sheltered home.

the sound of your own heartbeat or growling stomach. To help adjust to this new quiet world, the use of white-noise, such as a filter running in a fish pond or low-volume background music, will help greatly.

Ecostructure and the underground home

There seemed to be less tension on the part of my daughters once we had moved into our underground home. The whole feeling of increased awareness of our involvement with our surrounding ecostructure became apparent. The children seemed more aware of their need to be involved in helping others to become aware of their need to be more thoughtful towards the environment and the earth. I cannot explain this phenomenon, but I feel that it has something to do with the womb-like atmosphere created by an underground structure. The feeling of being a part of the earth itself tends to make me more aware of my need to not degrade any life form and help preserve the earth's ability to care for me. This is almost a spiritual feeling that I did not acquire until I had been living in my home for several months.

I have since visited with a number of people whom I have helped with planning their own underground homes, and they have reported a similar increased feeling of closeness to the earth and its creatures. There is definitely a secure and quieting effect created by an underground home that is missing in a surface home.

The interior

Since earth-sheltering is a most unforgiving type of building mode, too often too much time is spent in designing support systems, load-bearing walls and other necessary structural plans. In contrast, little consideration is given to the subject of interior use, and unhappiness with living conditions can result. The interior is what you will see (not the strength of the wall or roof support), and what you are surrounded with in this interior and how you are able to move about and use the space will determine your happiness with the dwelling.

Location of wet walls should, of course, be as centralized as possible, incorporating the double wall, back-to-back adaptation whenever possible. Care should be taken to have the water and electrical service enter and exit the structure as close to the location of wet walls as possible. Running pipe and conduit under permanent floor structures for great distances only increases the problems of repair should need for repair arise.

When determining the placement of wet walls, a working floor plan of the structure is essential (especially in areas that would involve plumbing) so that you can visualize the traffic patterns in order to avoid congestion, insure privacy and avoid unnecessary expense and duplication. At this point of planning, consideration of rough-in plumbing should be planned, since there is little chance to alter the location of service or make additions once the cement is poured. Rough-in expense is minimal, whereas later changes would be very costly and time-consuming.

If you are going to order new fixtures and appliances, it might be practical to select white as the color in order to make changing decorating schemes easier later on. If there are several members in your family, consideration should be given to using two or more "source" water-heating units, each to serve a specific area, rather than one large heater that will be more costly to operate. When teenagers are around, the cost for hot water really goes up. Clothes-washing areas and kitchen use are a good combination, with the other water-heating units divided to serve the children's bath and the master bath, if two are to be built.

Kitchen cabinets and utility-room cabinets can also be the same as those used in the bathrooms. Plan ironing-board storage, detergent storage, broom closets, sewing centers, a place to hang permanent-press clothing fresh from the dryer, and pantry storage at this point. Often a great deal of cost can be saved by ordering from one supplier, though I prefer salvage yards and yard sales, etc. If using prebuilt cabinets, be sure you select the hardware at the same time in order to avoid delays and needless frustration. Most lumberyards, mills and supply houses dealing in prebuilt cabinets will provide the sizes of cabinets needed if you give them accurate measurements.

If custom-built cabinets are utilized, be sure to have specific deadlines for the cabinet maker (in writing), and insure yourself against cost overrides. Open shelving and built-in cabinets that are built and installed on site are probably the most economical.

Plan the vent system of bathrooms, kitchen and utilities (hot water heater, clothes dryer, etc.) so that the exits of the vents are in a centralized location and as close to the sources of use as possible. Automatic thermostatically controlled vent switches are useful to prevent the inadvertent build-up of heat and moisture in windowless bathrooms. If ceramic tile is to be used in the bath or kitchen, choose a color that can easily be replaced. When choosing countertops, choose materials that will withstand heavy use and abuse (burns, heat, moisture). Check with the local mill for the cost of custom chopping-block as compared to Formica. The use of newer materials such as Corian will save you a lot of money, and, being more durable, it should never need to be replaced because of heavy wear.

Storage areas

Large storage areas, such as a storage wall with built-in shelves and closets, should be incorporated in each bedroom. The use of decks for children's rooms provides additional storage, and platform-bed pedestals with storage drawers are useful. Built-in window seats for seating, as well as storage of clothing or toys, would be helpful if there are any young children in the family. Plan all vents to each room to take advantage of the soil pipe and thermal chimney system. Use every available space for cabinets and storage. Since you will not have a basement or attic, it is important that you plan the storage areas ahead of time.

Bedrooms

When planning the bedrooms, the most important consideration would be the element of access. Rooms that open directly onto a general use and family area are not usually desirable, and considerable loss of privacy is inevitable. A central hall plan would be advisable. Consideration of light source is essential, for if rooms are closed off from a natural light source, the use of skylights or artificial lighting is necessary. All bedrooms are required to have a natural light source, so if they are located at the back of the structure, skylights are then mandatory. Visual planes are often lacking in completely enclosed rooms; therefore, mirrors, lighted aquariums and plant lights are all useful substitutes.

In the underground house, the use of the bedroom may tend to take on many more additional uses than in conventional structures. The use of sound-muting materials such as carpets, acoustical wall coverings, bookshelves full of books, the inclusion of television antenna jacks, etc., will insure your comfort in what will probably become your main place of solitude.

Living room floor plan

A complete wall of bookshelves in a living room is an important inclusion, and you probably cannot overplan the amount of wall space incorporated with shelving. A floor-to-ceiling dividing wall of bookshelves/storage cabinets would serve as both storage and room divisions especially if a very large area is being used. If the living area is open, which is common in an earth shelter, careful attention to the placement of entrances and exits is important. Be sure to include extra phone jacks and television antenna outlets at this point.

Living room lighting

Plan for adequate lighting that is also variable, using track lighting and ample wall outlets so that low-wattage area lighting may be used according to need. If the area is lit naturally from the greenhouse or atrium area, be careful to plan durable window coverings such as Levelor® blinds or wooden shutters to control both light and heat build-up. Plan skylights to the back of any area that fronts a glass wall. The area will receive a lot of light from the glass wall, but it needs to be balanced.

Defining space areas

In order to make the most of the living room and to give it a feeling of warmth, consider the use of moveable, or portable, walls. These may be on tracks which fold, or you can use smoked or colored sheets of Plexiglas suspended from chains to define the use of space. Changing levels by using conversation pits, decks or lofts can greatly enhance the feeling of space. When dealing with a single-level structure, variation of levels (if only by six inches) prevents feelings of spatial confinement.

Floors and sound

When choosing floor coverings (other than rammed-earth or adobe), the element of sound has to be considered. In a large area, sound will carry (and even echo), so the use of heavy rugs or carpeting is important. The floor covering should be durable and chosen for warmth of color to ensure a feeling of coziness. Wood floors are popular, easy to install, and will last the lifetime of the dwelling. Salvaged floors from houses that are being torn down cost a fraction of new wood and can be sanded when in place to look like new.

Greenhouses and atriums

Children bored during the winter with nothing to do? What better use of the attached greenhouse than for a giant sandbox or a garden where you can grow an unlimited variety of vegetables, herbs, decorative plants or fish. At the same time, your creation aids in the storage of heat for the entire underground house. The working greenhouse can be used in both summer and winter.

The use of a greenhouse or atrium in conjunction with the design of the earth-sheltered home will provide a source of heat and natural light to the structure. When designing and planning the use of the greenhouse or atrium, it should first be decided if the greenhouse/atrium will be a working structure or a source of light and decoration only. It seems that a great many people overlook the many possibilities that can be incorporated into their total living environment when they bypass the use of the greenhouse/atrium as an integral part of the structure. Plan the greenhouse/atrium carefully so that it will store the heat it gains on a sunny day to heat the structure during the night. Using a gravel bed or incorporating several tons of rock in storage under the floor will add to its heat storage capacity. The use of a brick floor over the storage pit area will increase storage capacity and be attractive as well.

The greenhouse/atrium, if carefully preplanned, can also be used for the ultimate in recreational rooms. The installation of a hot tub is perhaps the best single addition, being both recreational and energy-efficient. The tub would be heated by the passive gain from the glass during the day through a thermal syphon system, and would act as an additional heat sink in the evening.

Water as a recreational source

The fish pond serves in somewhat the same capacity as the hot tub, in addition to serving as a source of food. Catfish raised in a pond do well in the warmth of the greenhouse, and their waste can be recycled to the plants. The plants in turn furnish additional food and convert your carbon dioxide back into clean oxygen. Even the installation of a swimming pool is possible in a large greenhouse.

Whatever recreational additions may be incorporated in the greenhouse/atrium, the very careful planning of water service and drainage is imperative. Much of the enjoyment of watering your indoor garden will be lost if you find it necessary to carry water to the garden from a distance. Generally speaking, a hose should not have to be longer than thirty feet to meet watering needs. Inside the structure, distance should be even smaller to enable you to use the water where it is needed without the hassle of an overly long hose.

Temperature control in the greenhouse/atrium

The greenhouse/atrium addition, as previously discussed, is incorporated into the earth-sheltered structure to provide heat to the structure in cold weather through direct passive solar gain. What about summer? Careful preplanning to protect and shade the glass is essential in order to prevent a most uncomfortable temperature gain during warm and hot weather. There are several ways of managing this gain problem, the best of which is the use of an overhang on the greenhouse which will, if correctly calculated as to the angle of the sun and the magnetic position of the structure, all but eliminate the gain during warm weather.

Over the long run, the additional application of exterior movable shutters would be a good investment. These shutters could shade, filter the light and protect the glass during heavy wind and hail.

Thermostatically controlled vents for a greenhouse are essential if it is to be a working greenhouse. I would mount two solar chimneys on the roof as part of the draw for the soil-pipe system. This makes the sun work to cool, rather than heat, the greenhouse. In the winter these chimneys are dampered down to allow a modest air exchange.

The loss of heat through the greenhouse glass during winter nights is not as great as most people imagine it to be, though it does occur. The use of insulating curtains, custom-cut blinds, insulated removable panels or heavy drapes will reduce this loss. If the greenhouse glass is well protected from the cool prevailing winds, then there is no windchill on the glass, and you are losing less heat to the ambient air temperature. I did not use any glass covering on my underground home in the winter since I felt what little heat loss occurred was more than compensated for by the BTUs collected in the mass of the structure.

This underground house near Emerald, Nebraska, uses a surface structure to house the garage and provide additional heat and light to the back of the underground structure.

Interior finish of a greenhouse

When planning the interior finish of the greenhouse, thought should be given to using materials that will withstand the conditions to be imposed upon them such as moisture, the spray of water from hoses, the destructive ultraviolet rays of the sun and general hard use. Wood is the best material since it is impervious to most of the changes that will occur during the seasons. Being durable, the wood will withstand hard use. It will not mould, rot, discolor, crack, peel or chip. Wood will only age and become more attractive.

Compositions should be avoided since they will break down and separate after only a few years. Plastics are available and are fairly resistant to moisture, ultraviolet rays and hard use. They will not weather like wood and will eventually succumb to scratches. Because of their nonporosity, plastics encourage the growth of mould and mildew, presenting a cleaning problem. Many plastics will emit unpleasant and even toxic fumes when exposed to great amounts of ultraviolet light and/or heat. The levels would probably not be high enough to injure you, but they will certainly inhibit plant and vegetation growth. Also, in event of a fire, many of these plastics give off deadly cyanide gas.

Underground garages

Garage use in the underground home will be different than in conventional structures. During the initial planning stages of the structure, allow for a larger garage area since it can be used for many other things other than the storage of an automobile. Plan the entrance to the garage from the inside of the structure so that it is not in the main area of traffic and so it preferably opens into the utility area (easier to keep the floors clean that way!). Include a floor drain and water service in the garage. Electrical service to the garage will be necessary to provide for general lighting and outlets for tools.

I loved my garage in my first underground home. The winters in Nebraska are severe, but since the temperature in an underground garage is constant, my old car would start regardless of the weather. Only two problems were encountered with my garage. The warm temperature was just perfect to speed up the rusting on my car that the salt on the iced roads had already begun, and the floor drain was necessary in order to wash off the salt. The other problem that I encountered was that of drifting snow in front of the garage entrance. Due to the fact that the garage faced south, along with the greenhouse and porch, the wind would whip snow over the roof and pile it neatly by the entrance in five- to ten-foot high drifts. I never did find an easy solution to that one—I had to keep my snow shovel at the ready in order to be able to create a path to the car. Many people with surface homes and heated garages would also experience similar problems, the main difference being that I did not have to pay for the heat that warmed my garage.

23
Commercial Applications

Building underground is an alternative to traditional above-surface structures that should seriously be considered by businesses, since to build one would result in considerable cost savings. The more simple its design and construction, the less maintenance, mechanical heating and air-moving equipment the structure will need. Most notable, however, is the fact that the structure will cost much less to construct than a traditional above-ground building.

An underground manufacturing plant

The Trump Mfg. Co. in Plattsmouth, Nebraska, is a good example of how a business can benefit by building underground. Trump, a relatively small company located in a small rural town, manufactures aircraft de-icing and washing equipment for large aircraft utilized by airlines and cargo lines. Trump employees fabricate these large pieces of equipment and mount them on trucks inside a twenty-thousand-square-foot plant that has a ceiling height of over thirty feet. The structure is so strong that it supports a huge crane from the Twin Tee roof sections, while also supporting two feet of sod on the roof.

The plant is built into the side of a steep hill next to a main highway, but it cannot be seen from the highway itself. The downhill side of the structure has a number of large bays from which vehicles are driven.

The manufacturing area is heated above the 55-degree-Fahrenheit ground temperature in winter by the heat generated from the welders and other fabricating equipment.

The design of the structure is quite simple. It is a large rectangle that utilizes staged, poured concrete walls with steel rod reinforcing and prefabricated Twin

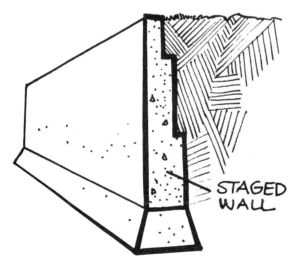

A staged wall (thicker at its bottom and thinner at the top), compensates for lateral earth pressures incurred by structures with high exterior walls.

Tee roof units, topped with a two-inch concrete cap. The structure was insulated on the top and top side-walls. The entire surface of the structure was waterproofed and then backfilled after a drain field was established.

As mentioned earlier, the deeper the excavation, the more pressure that will be multiplied at the lower portion of the walls. In order to avoid stub walls and other obtrusions on the inside of the structure to reinforce the walls, the walls were made thicker the deeper they were. The walls were formed in stages.

Twin Tees can be designed in various sizes to span great distances and carry heavy loads, while allowing rapid completion of the structure without all of the scaffolding and internal bracing that a poured-in-place deck would require. The illustration shows how a Twin Tee functions. Another real advantage to Twin Tees is that they can also have openings cut in the deck between the stems of the Tees without causing any weakening.

Trump cut round holes in the deck at the rear of the structure and set precast three-foot-diameter drain pipes in them for ventilation and auxiliary lighting. Gases and heat are drawn through the structure and vented naturally through these openings. The drain pipes serve to cool the structure in the summer and to vent in the winter. The Trump building can be built at about equal cost to a steel prefabricated structure, but the energy savings and maintenance savings should pay for the construction costs in no time.

Schools

Several rural school districts have voiced an interest in my school design (illustrated previously under arch design) because of simplicity, ease of construction and low cost. The energy savings of this design will easily pay for the structure in a ten-year period, based on current costs. The modular concept, the amount of light and the functional space also make it an excellent design for commercial offices. The central dome would be used for cafeteria space, meeting areas and other business functions that require space for large audiences. Mutual of Omaha's three-story underground facility has a central dome that serves as a cafeteria and side meeting rooms.

Shopping mall

My commercial, institutional, agricultural and factory designs are all relatively simple, inspired by simplicity and functional open floor plans. Adding tunnels or arches to the wheel design enlarges the space, and the spoke tunnels meet all codes and allow light and egress from both ends as well as through the light wells. This design would make an exceptional shopping mall for retail space in medium-sized communities of around 15,000 to 25,000 people.

Either the steel-culvert or the bent-rod method of forming the arch works well on this design and is not difficult to construct. The interior finish can be as elaborate as the client desires. The clear-span arches are easy to upgrade on interior finish at a later date if desired. None of the interior walls are support-structures; therefore, these walls may be removed or changed at any time to meet increased space demands or expansion.

Mobile-home earth shelter

Another economic office space for a business requiring a small amount of space (such as an art agency, ad agency, small restaurant or any small retail outlet) would be the mobile-home earth-shelter concept. Modular and mobile homes can be

Underground tunnelled space in Kansas City, Missouri, provides office storage and manufacturing space. The entryway to the structure is shown.

purchased with about any floor plan desired. Inserted into a landscaped, post-and-beam shelter, it will become an energy-efficient commercial establishment. One big advantage to this structure is that no one will guess that it is a mobile home (if it is adapted correctly). Since the mobile home is already equipped with plumbing, fixtures, cabinets, lighting and other details that account for over half of the construction costs in most buildings, you save time as well as money.

Quarries

Almost all of my home designs have possible commercial applications, the Convertible Crescent being especially adaptable to commercial, institutional or retail use.

One possibility that many overlook is abandoned limestone quarry caves. The Kansas City area has many abandoned quarries. Kansas City has converted over 15,000,000 square feet of this underground space to commercial use. Everything from storage or warehousing to retail space and offices has been created there. Considerable savings in energy have been gained. The preexistent space needed only nonsupporting dividers and other basic (but easily installed) fixtures to make the space usable at a fraction of the cost of surface construction.

There are many areas in the country with mined space such as this that can be utilized. If you are contemplating use of space such as this, obtain the services of a competent mining engineer. Many times the capstone has been removed in the mining operation, making the space unsafe and damp. The mining engineer will do a complete survey of the excavation and list all of the possibilities.

Bunkers

Another example of a preexistent underground space is the many ammunition bunkers that the United States government had built around the country during World War Two. Many of these bunkers have been successfully converted to factories and other businesses at Hastings, Nebraska. This government property, which has been idle, can be bought for very little and converted to profitable enterprise. Hastings has several hundred of these structures, many of which have been converted into an industrial park. An enterprising builder, contractor or developer could reap considerable profit with little cash outlay by converting such structures. Old Nike missile bases offer another opportunity for commercial conversion. If you are a contractor or businessman who is interested in such underground conversions, contact the United States Government for more information.

Agricultural applications

Underground structures are used extensively in Europe for agricultural purposes. Hog-farrowing operations and poultry operations are especially sensitive to radical temperature changes, so the stability of underground habitats becomes very attractive. The simple A-Frame farrowing house has been used successfully and is very economical to build. The structure can be built in the farm shop and transported to site for backfill. The use of a soil-pipe ventilation, dehumidification and cooling system attached to this structure makes the environment as near perfect as possible for farrowing. The building is contructed of 2 by 6s and sheeted over with three-quarter-inch marine plywood. Two sheets of six-mil polyurethane serve as a moisture barrier. The hogs' body temperatures

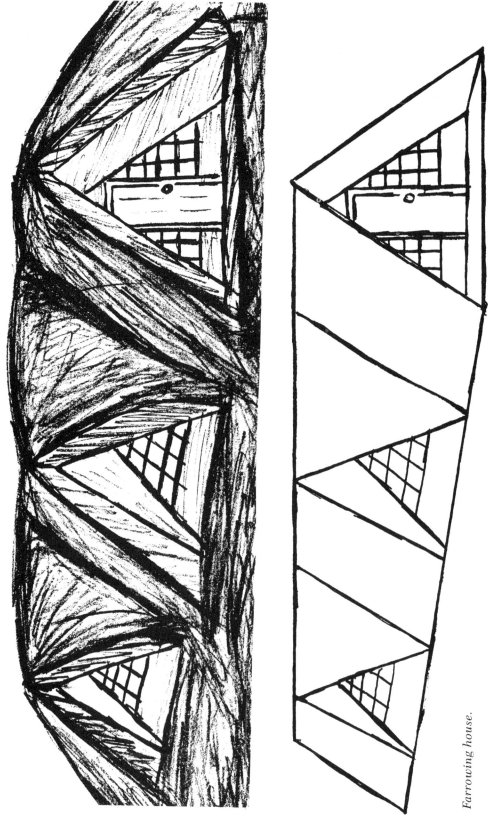

Farrowing house.

maintain all of the heat that is necessary in the winter and the soil pipe and thermal chimneys provide the necessary cooling and dehumidification in the summer.

A center aisle allows the operator to feed the animals and clean the structure on either side with ease, while giving the sow the correct amount of room. Farrowing crates or pens fit well. Propane is becoming increasingly more costly, so this structure should be considered in colder climates where hog operations require propane heaters to cut losses in winter. Properly maintained, these farrowing houses should last almost indefinitely.

Brooder houses and laying houses for poultry are another ideal application for earth sheltering. Temperature fluctuations tend to erode the productivity of a laying operation. The brooder house is a simple post-and-beam structure constructed in the same method as the Trench House. Posts are set with girders spanning between them, beams are on four-foot centers, and rafters are placed on two-foot centers with sheeting over the frame. Polyethylene covers the entire structure, which is backfilled as desired. A floor may be poured or built on sills as a wooden floor. Soil pipe and thermal chimneys will cool and ventilate in the summer, as well as dehumidify. These are rather simple structures and may not appeal to large operations. If that is the case, cement block or poured structures may be substituted. Many agricultural applications (such as dairy barns and milking parlors) would also benefit from earth sheltering. The main incentive will be the savings against the continual rise in fuel and energy costs where heating of a structure is needed.

Post-and-beam chicken house utilizes cordwood fill between posts at the front of the structure and wood or railroad-tie shoring on the side, back and retainer walls.

Root cellar

Root cellars have long been a dependable method of preserving garden produce. The root cellar, abandoned by most truck-gardening operations long ago in favor of refrigeration, is now on the rise once again. A good friend of mine has a large orchard operation, but he cannot afford the refrigeration equipment that would be necessary to handle six thousand bushels of apples. Although refrigeration will reduce the temperature of the apples below that of a root cellar, the root cellar will easily store the apples over a three-month period. We have designed a tunnelled space that will handle the quantity of apples expected from the orchard. It will be located within one hundred feet of the orchard and will incorporate a loading dock as a part of the tunnelled space. Since the space will require no finishing, the only cost will be the labor used to dig the space. The building of an on-surface refrigerated unit of equal size would cost up to ten times as much as the tunnelled space. If tunnelling conditions are not good, the arch formed by inserting rebar into a footing or even directly into the ground, then bending them to shape, tying them and covering with mesh and pouring a concrete skin would form the same space at still a very reasonable cost. The poured arch would then be backfilled and ventilated with soil pipe and thermal chimneys. The tunnelled space should also be ventilated with soil pipe and chimneys. Any produce storage must be dry and dark, have a top and bottom ventilation, and maintain a constant cool temperature. Either the tunnelled root cellar or the poured and backfilled space will meet these qualifications.

Andalusian earth shelters

Numerous underground villages can be found in Andalusia, Spain. The foothills of the Sierra Nevada mountains offered remote but curiously beautiful sites for dwellings carved into the soft tufa rock.

Underground homes in Sacramonte, Spain.

At Sacramonte, near Granada, the whole southern hillside is covered with whitewashed caves. The doorways are decorated simply with clay tile projecting over them. Many doorways have glass bead curtains to allow colored sunlight and fresh air to enter the home while still maintaining privacy. Over a dozen underground villages dot the countryside. The town of Guadix has several thousand underground dwellers.

While the shelters oddly blend with the unusual cliffs, their white-washed chimneys and retaining walls stand out. The retaining walls act as barriers to excessive rainwater, directing the water to spillways at the ends of the building. On numerous occasions, the courtyards are covered with vines during the summer to provide cool outdoor living space. During the winter, the huge retaining walls reflect and concentrate sunlight to brighten the interiors. The unusual chimneys vent smoke during the heating season and act as fresh air ventilators to distribute fresh cool air through the entire structure during the summer.

The basic dwellings are simple, even crude, but complex shelters with as many as twenty interconnected rooms can also be found here. The homes are equipped with lights, television and running water. In some villages craft shops line the main road, dug into the banks, and even bars can be found underground.

A craft shop in Andalusia, Spain.

Exterior of Mutual of Omaha Insurance Company's huge underground office complex in Omaha, Nebraska. The dome over the cafeteria is all that appears above ground.

Mutual of Omaha underground offices

A unique office building in the city of Omaha, Nebraska, is the three-level underground addition to the Mutual of Omaha Company's International Headquarters. This underground structure adds nearly 190,000 square feet to the complex.

A massive glass dome caps the center of the building, covering the 1200-seat cafeteria. This dome ranked as the largest of its kind in the United States at the time of its construction. The dome is 90 feet in diameter and has a rise of 15 feet. It contains 504 pieces of one-quarter-inch, bronze heat-tempered glass and 504 pieces of clear heat-tempered glass. The top-center curved piece is the only plastic material in the dome. The dome has over 30 tons of material in its construction. Two Omaha-based firms, Peter Kiewit Sons, Inc., and Leo A. Daly Company, served as general contractor and architect, respectively. Work began in January 1978 and was completed in early 1980.

About a two-thirds saving in heating and cooling costs is being realized over what would be projected for a surface structure of equal size. The building cost about $5 million less than it would have cost to build a comparable above-ground facility featuring the same materials as found on the existing Home Office building.

Work on the excavation took nearly nine months; 150,000 cubic yards of soil (over 10,000 truck loads) were removed from the site. When completed, the hole measured 44 feet deep, 260 feet wide and 285 feet long. This was the most extensive underground project undertaken in the Omaha area since the Strategic Air Command built its underground control center in 1955.

The first level (top floor) of the underground structure includes an employee cafeteria, kitchen, employee library and lounge, company archives and an employee training center. A sunken garden court featuring a water fountain and fig trees is located directly underneath the dome. The second and third levels provide some office space, though they are primarily used for storage of office records. An all-electric filing system is the focal point of the third level.

The area surrounding the dome and covering the top of the building is landscaped. Three emergency exits with skylights are located on three sides of the dome. The dome is lit at night.

DESIGN DETAILS
MUTUAL OF OMAHA UNDERGROUND OFFICES

GENERAL
Area: 184,000 sq. ft.
Site area: 75,000 sq. ft.
Number of floors: 3 levels

Types of space: First level: 1,000-seat cafeteria, archives, library, lounge, training and development conference center; second and third levels: office space and storage.

CONSTRUCTION
Dome: 90-foot-diameter glass double-glazed; rises 15 feet above grade to top 45-foot-deep building; two layers of tempered glass separated by ½ inch of dead air space; outside glass solar bronze; inside clear; U value of 0.55.

Walls: Concrete encased in hot applied liquid-rubber base waterproof membrane.

Floors: Metal deck and concrete fill.

Structural system: Steel beams and girders.

Roof: Structural concrete roof deck; surface received liquid-applied rubberized asphalt waterproofing membrane, finished ¼ inch protection board, 3 inches of crushed rock, 3 inches of high density styrofoam insulation, 1½ inches of crushed rock, a permeable membrane, 12 to 18 inches of manufactured topsoil, sod.

HVAC DESIGN
Heating: Primary mechanical equipment not included since building will utilize company's central chilled water and steam plant; medium velocity air to variable volume induction boxes; low velocity air from box light troffers.

Normal metric degree days: 3,673

Winter design conditions:
Temperature of ground, 50 degrees F.   Heating MBH: 473.
Outdoor −6 degrees C (−10.8 degrees F)
Indoor 20 degrees C (68 degrees F)

Summer design conditions:
Cooling tons: 557
Outdoor (95 degrees F)   (77 degrees F)
         35 degrees C Dry Bulb   25 degrees C Wet Bulb
Indoor
         (78.8 degrees F)   (66.2 degrees F)
         26 degrees C Dry Bulb   19 degrees C Wet Bulb

LIGHTING
Type: 2-lamp fluorescent, parabolic louvers

Intensity: 70 footcandles general illumination. First level on dimming system.

CREDITS:
Owner: United of Omaha
Architect/Engineer: Leo A. Daly
Foundation Consultant: Woodward-Clyde Consultants
Audio-Visual Consultants: Jamieson and Associates
Contractor: Peter Kiewit Sons' Co.

University of Minnesota: Williamson Hall

One underground building which has been widely publicized is Williamson Hall at the University of Minnesota. Another structure, over 100 feet below the surface, will be built under the campus in addition to the bookstore in Williamson

Hall. The new 110,000-square-foot structure and Williamson Hall were designed by the Minneapolis architectural firm of Myers and Bennett. This firm is heavily involved in the design of underground structures for commercial and educational use.

Williamson Hall houses the University Bookstore as well as the Admissions and Student Aid offices. The structure is built on four levels below ground plus a sub-level for mechanical equipment. Four overhanging garden terraces are spaced at each level on the inclined south- and west-facing portions of the structure. These terraces allow light and heat gain in the winter while shading the glazing in the summer and allowing indirect light into the structure during this period.

A concentrating collector array on the roof consists of ten curved mirrors, 12 inches wide by 20 ft. long. Each mirror focuses the sun's rays onto a single copper tube. Hot water provides heat for the structure when needed and absorption air conditioning in the summer. Just five percent of the floor area is above ground. This five percent accounts for 34 percent of the heat loss from the entire building and is a good illustration of building underground using as little surface exposure as possible.

Williamson Hall, University of Minnesota.

The multi-level construction of Williamson Hall is shown here, along with overhanging garden terraces.

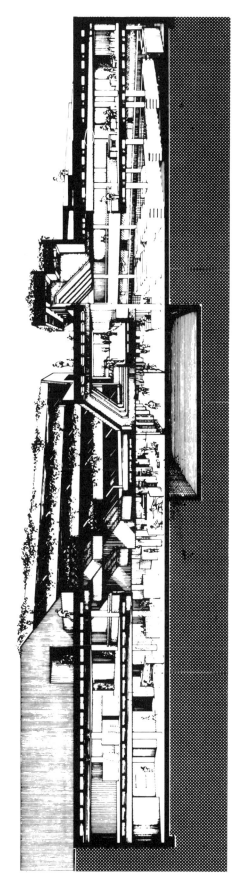

Floor plan of Williamson Hall.

The estimated cost of site work for this structure, which included clearance, excavation and hauling, sheet piling, backfilling and subsurface drainage tile, was less than the typical costs to face exterior walls of a surface building of this type and size. This 83,000-square-foot building was completed in 1976 for a cost of $3,500,000 and has been monitored by instruments on a grant from the Science Foundation's Energy Conservation program. The data certainly has been encouraging since additional underground construction will be the pattern for the University building program.

Civil/mineral engineering building

This structure, presently under construction, will also be 95 percent underground. As designed by the architectural firm of BRW Architects, it will have 142,300 square feet of space that is to be utilized for classrooms, laboratories and offices. It will include a bus terminal along the northern edge of the site, between the underground laboratories and the Field House (here the buses will create the least amount of environmental disturbance).

The underground classrooms are arranged in a series of fan-shaped segments in 30-degree intervals, stepped down successively so that their roofs form a great 150-degree spiral entry court which begins at grade level and terminates in the main entry to the building.

The "active systems," including solar heating, solar electricity generation and ice energy cooling, will not only reduce the need for conventional energy sources, but will serve as an education to the engineering students enrolled in the school. The basic principle of the system is that a high-temperature solar collector will be used to heat the building and to partially electrify it through a cycle generator. Summer cooling is achieved by using ice that is stored in an underground tank, formed and employed by a system of heat exchangers at the surface of the building's cooling system.

"Passive systems" include earth sheltering and underground space, landscape microclimatology and passive solar heating (employing hybrid systems). Landscape microclimatology is another way of saying that vines, deciduous trees and shrubs are used strategically to shade and cool in the summer, while allowing sun and heat to come through in the winter. The hybrid solar systems include a water-filled trombe wall that is connected to the fresh air supply to the building so that this air is heated in the winter. This is combined with an exhaust air-heat recovery system and will supply a significant amount of the heat required for the mined space of the structure.

A solar optical system, using a series of lenses, mirrors and reflective surfaces, will gather and concentrate sunlight and direct it through the interior of the building. As environmental design, it has the potential of enhancing the habitability of deep interior spaces by distributing sunlight to these areas. As an energy design, it promises to be the most cost-effective of all since it will provide about 10,000 foot candles of light. It transmits and spreads the light using no energy other than its own.

The construction involves not only cut-and-cover excavation techniques, but also mining of deep earth space. This building is not only cost-effective, it is also a living lesson to the students that will inhabit the structure.

Appendices

Glossary

ADVANCE
    Digging progress from the entrance inward.

ALLUVIUM
    Any deposit such as sand or mud that has been transported and deposited by flowing water.

ALLUVIAL
    Composed of alluvium and various clays along rivers.

ARCH
    A curved structure spanning an opening; a bowl-like curve.

BERM
    An artificial mound of earth. The structure is generally built above grade, and a soil cover is then sculpted onto the structure to appear as a mound (said to be bermed).

BOND BEAM
    A beam that is usually cast horizontally along the top of a block wall and tied to it by reinforcing steel bars. This beam then provides lateral stability to the wall.

BOX BEAM
    A poured box-shaped beam that is integral with the roof slab.

CAMBER
    A unit of material having a slight bend upward.

CAPILLARY DRAW
    The transmission of moisture from molecule to molecule.

CENTER PENETRATIONAL
    Refers to an earth-sheltered structure that has its natural light and entry in a center courtyard.

CHASE
    A groove or channel in a floor that wiring or plumbing can be placed in and covered with a removable panel, allowing easy access.

CHINKING
    A method of placing mud, straw, daub or cement between logs to limit air infiltration.

CLAY
    A fine-textured earth that is plastic when wet, but hard and compact when dry; used for ceramic ware, brick, etc.

COLD JOINT
    A cement pour that occurs in two phases whereby one pour sets up before the other, thus creating a joint.

COMPRESSION
   The pressure exerted by the weight of the soil.

CONDUIT
   Pipe that encloses electrical wiring.

CORDWOOD CONSTRUCTION
   Often referred to as stackwood and firewood. It is both a material and a method of construction. Pieces of cut wood between sixteen and twenty-four inches long are stacked (similar to a stack of firewood) to form a wall. The pieces are mortared on both ends with a piece of insulation between to provide a thermal break. Complete structures can be built using this method, or this method can be combined with post-and-beam modes of building.

CORES
   Hollow spaces cast in any concrete unit, such as cement block, Flexicore, etc.

COVENANTS
   Rules or regulations drawn up by people comprising a neighborhood.

CUT AND COVER
   An open excavation that is covered back over after a building has been erected in the excavation.

DEADMAN ANCHOR
   An object that forms an anchor, such as a cement block or a cement ball, that has a cable or rod embedded in it and is attached to a wall or post. The anchor is then buried, using the weight of the soil to immobilize the anchor and thus stay the wall.

DEW POINT
   The temperature at which water vapor in the air attains a condition of supersaturation with respect to relative humidity, giving off moisture (as in the case of a cool drink in summer which "sweats").

DOUBLE ENVELOPE
   Essentially two structural shells, one inside the other. Usually heat, air and ventilation of the inside living shell are introduced by using the space between the two shells. This space also acts as an air lock to further insulate the inside living space.

EARTH PIPES
   Pipes or ducts located in the earth that temper air temperatures being introduced into a structure. Pipes are usually buried below frost lines.

EARTH STABILIZATION
   A method of coating and sealing the exposed soil surface to keep it from becoming damaged or eroded.

EXPANSIVE PRESSURE
   The pressure exerted by soil once it has been saturated or frozen. This pressure can affect the wall that is backfilled, both on vertical shear and horizontal shear.

EXTRADOS
   The exterior arc of an arch, such as in a tunnel.

FACE
: Vertical bank or cut into which one tunnels.

FIRE SETTING
: Process of cracking or spalling a working face with fire.

FLANGE BEAM
: A steel beam that has a vertical center section and a flange on the top and bottom. This beam looks like a capital letter "I" when viewed from either end.

FLEXICORE
: A trade or brand name for a concrete precast and prestressed unit that is used in series to form a ceiling or floor of a structure. These units usually are two feet wide and up to twelve inches thick. They are cast in lengths up to sixty feet and cut to the length the customer orders. The units are cast out of concrete with a heavy aggregate mix. They usually have three hollow cores running their length and have either four or six steel cables stretched through their bottom half to stress the unit.

FLYING BEAM
: Usually an enlarged portion of a poured roof section that forms a beam to support that particular section of the roof. Steel bars are usually tied inside for shearing strength.

FRENCH DRAIN
: A gravel field placed around a structure to allow percolation and drainage.

FULLER'S EARTH
: An earthy substance, lacking plasticity, resembling potter's clay.

FULMINATE
: A salt of highly explosive fulminic acid.

GEOTHERMIC (GEOTHERMAL) GRADIENT
: Increase of the temperature of the earth from the surface downward, averaging about one degree Fahrenheit for each sixty feet.

GEOTECTURE
: Subterranean construction which usually involves utilizing mining and tunnelling techniques.

GNEISS
: A laminated or foliated metamorphic rock corresponding in composition to granite.

GRADE
: The level, incline or decline of a floor.

GRADE; GRADED
: To determine the rate of increase or decrease of a slope; sloped upwards or downwards.

GREY WATER
: Water that has been used in the household (dishwater, lavatory, shower and tub water) that does not contain solids.

GROTTO
: Small cave or vault.

GROUT
: To fill out for finish with mortar or cement.

HEADING
: The end of a drift, gallery or tunnel.

INTRADOS
: The interior curve of an arch, such as a tunnel lining.

JOIST BEAM
: One of several beams or supports that are parallel to each other and support a floor or ceiling. They can be poured monolithically out of concrete when the floor or ceiling is poured.

JUMPER
: A steel bar used in manual drilling.

KEYWAY
: A channel between two precast members that has a bar of reinforcing steel inserted into it and is then filled with cement to tie the units together.

KICKER PLATE
: Any ninety-degree angle of material that prevents a reinforcing brace from kicking outward when coming under load.

KNEE BRACE
: A brace placed at an angle between a post and its overhead beam or girders. It is designed to shorten the span of the beam or girder and change its point of shear.

LAY UP
: To place bricks, rocks or other building material into a wall, using mortar.

LEACH FIELD
: A filtration bed through which grey water is passed before being used in gardens or lawns. This bed could be a field of gravel or charcoal.

LOAD
: Weight of soil imposed upon a vault.

MARL
: Crumbling deposit consisting of clay and calcium carbonate.

MATTOCK
: A tool for digging and grubbing which resembles a pickax but has blades instead of points.

MORTAR
: Any compound used to level and adhere various building materials together (usually cement to bond bricks or cement blocks into a wall).

MUCK; MUCKER
: Broken rock or other material that is to be removed; the worker occupied with loading ore or debris.

OOLITE
: Limestone rock of the Jurassic system (a geological period which occurred approximately twenty million years ago), that consists of small round grains resembling fish roe that has been cemented together.

PAYBACK
: The time period anticipated to save enough money to justify the installation of labor- or energy-saving appliances, products or designs.

PERCOLATION
: The indication of the porosity of a soil which allows moisture to penetrate it and to percolate or dispense through the soil as opposed to being trapped on the surface.

PHYLLITE
: A crumbling slaty rock metamorphosed by chemical and mechanical action under great pressure.

PILASTRE
: A supporting column that receives a compression load.

POST-AND-BEAM CONSTRUCTION
: A method of construction whereby the structure is supported by a series of posts set into the ground that support horizontal beams or girders, which in turn support rafters and sheeting for the roof. The structure then gets its support from these few posts and beams as opposed to a stud wall built of many smaller vertical and horizontal pieces of two-by-fours.

POST TENSIONING
: Tension applied to a cement unit. The tension is supplied by threaded rods or exterior cables after the unit has hardened. These rods or cables are then permanent, exterior tensioning devices. The process is similar to placing your hand on both ends of a row of books, pressing inward, and then picking the books up as a unit.

POURED-IN-PLACE ROOF
: A formed roof into which ready-mixed concrete is poured.

PRESTRESSED
: The tensioning of a precast concrete unit. The tension is usually supplied by cables stretched through a form that is poured. When the cement is set, the cables are released, compressing the concrete unit.

PYRITES
: Common minerals of brass-yellow color, popularly known as "fool's gold," containing iron, copper or tin.

RAKE
: A timber placed at an angle.

RAMMED EARTH
: Both a material and method of constructing walls for a structure. The material contains seventy percent sand and thirty percent clay or alluvials, usually with a mix of one shovel of cement to seventeen shovels of the above mix. This material is then tamped or rammed in a heavy form until it will not compact further. After curing, it is a durable wall that is economical to build. The material should be moist enough to make a ball when squeezed in the hand, but still break and crumble when dropped on a hard surface.

REBAR
: Reinforcing steel bars or rods of varying diameters and lengths which are specifically manufactured with bumps or ridges along their surfaces. When cast in a concrete form, these rods keep the cement from breaking away or shearing, thereby supplying greater strength.

RETURNS
: Short or stub walls that are at ninety-degree angles to any other wall. These are stabilizing walls, usually found at the front of a structure, which support side walls.

SCHIST
: A metamorphic crystalline rock with a closely foliated structure such as mica and hornblende schists.

SHEAR
: The point at which a girder beam, rafter post or any tensioned building member will break.

SHEAR WALL
: A butting wall that supports another wall at right angles.

SHOULDER
: The place where the arched roof of a tunnelled space begins.

SIDE PENETRATIONAL
: Refers to an earth-sheltered structure that has its natural light and entry open to the side, generally in order to obtain a specific view.

SINTER
: To fire and fuse a mineral without actually melting it.

SLOPE SHIFT
: The movement of soil down the face of a slope.

SOFFITS
: The overhang or eaves on a roof system.

SOIL TYPES
: Sand, clay, decomposed granite, rock, gravel, shale, ledge loam, silt, expansive clay, etc.

SOIL PIPES
: Any pipe buried beneath the surface that is used for ventilation and dehumidification.

SOLAR CHIMNEY
: Any flue-like structure that can be heated by the sun to create an updraft, which will move air in a structure.

SPALLING
: To break off small flat pieces of rock, historically by means of fire-setting, presently accomplished by the use of explosives. Rock under excessive tension may also spall.

SPAN
: The ceiling between two support walls or columns.

SURFACE BONDING
: A term used to describe the process whereby cement blocks are dry-stacked and then coated with a cement material that has been impregnated with fibreglass strands.

TAMPING
: A method of packing or compressing soil by walking over it or by packing it with a weighted object.

TENSION
: The stress or pressure exerted on a building member.

TERRATECTURE
: Near-surface building, usually by the cut-and-cover method.

THRUST
: A reactive force. Example: the earth pushing on the bottom of a basement wall, or any force of a load moving horizontally.

TILT-UP OR TIP-UP
: A term referring to wall units that can be cast on the ground next to a footing and then tipped or tilted up on the foundation next to each other to form a wall.

TROMBE WALL
: A wall that absorbs heat and then radiates it back when there is no sunlight.

TRUSS
: These are constructed beams that support various imposed loads such as a roof, ceiling, floor, bridge, etc. A truss is usually a two-beam unit with braces constructed between the beams to add reinforcement. A truss can take many shapes and is engineered according to the loads that will be imposed.

TUNNELLING
: Process of digging an underground passageway or space.

TWIN TEE
: A tradename for a prestressed concrete member poured in a form with steel cables under tension. When viewed from the end, the member looks like two capital t's joined together at the top of the "T."

VAPOR BARRIER
: Usually a sheet of plastic (such as polyethylene) that is placed between two permeable materials to stop transmission of fluid and gases between them.

VAULT
: An arched roof or ceiling.

WATER TABLE
: The upper area of the part of ground that is completely water-saturated.

WET WALLS
: Any hollow core wall that is used to carry pipes and connections for toilets, sinks, lavatories, tubs and showers. It is best to use one of these walls to hold all plumbing, thus reducing the length of pipe runs and making it easier to reach and repair breaks.

Preplanning Checklist

1. Are you country material?
2. Site selection/total building environment
   - Legal description
   - Political control
   - Schools
   - History
   - Neighbors
   - Soil composition
   - Water
   - Drainage
   - Shopping
   - Utilities
   - Recreation
   - Access
   - Labor–materials market (indigenous materials, recycled, new)
   - Landscaping
3. Involve the family in planning the structure, including determination of size, space perception and design choices, and in the mortgage and money planning process.

Builder's Checklist

1. Survey for excavation.
2. Determine types of excavation equipment to be used, and select excavation contractors (i.e., front end digging, hand digging, backhoe, dragline).
3. Dig footing, drain tile and soil pipe trenches. (If post-and-beam construction, prepare and set posts.)
4. Pour prepared footing or post piers. Place drain tile and soil pipe. If a separate plenum is required for a soil pipe, it should be put in place at this stage.
5. Prepare and dig chases for plumbing, sewer, electricity, gas, telephone, cablevision.
6. Erect shell walls; i.e., cordwood, rammed earth, concrete, cement blocks, logs or shoring.
7. Install roofing system; i.e., poured-in-place roof; girder, beam, rafter and sheeting; Flexicore or Twin Tees; truss and sheeting. Be sure all openings for skylights, flues, vents, etc., are made.
8. Water-vapor barriers, insulation, drain field and backfill, including earth-cover for roof, should be taken care of at this point.
9. Prepare for and construct structural appendages (greenhouse, etc.). Be sure all footings exposed to freezing conditions are below the frost line and are insulated.
10. Begin construction of dividing walls, wet walls and various level changes. Run necessary conduit, plumbing, etc.
11. Prepare floor surfaces.
12. Enclose structure, including glazing of surfaces, doors, etc.
13. Complete interior finish.
14. Landscape.
15. Move in and enjoy. You earned it!

INDEX

Adobe homes, 25
  floors, 69
A-frame; see Deep Woods A-Frame.
Air management, 67–68
Air movement
  double dome, 42–43, 44
Andalusia
  underground villages, 186–187
Arches
  building, 145–146
  tunnelling shelters, 160–161
  Wondrous Warren, 154, 157
Arch system, 145–148
Atrium, 176–177
Bathroom
  basic information, 174–175
  Prairie Cocoon, 66
  toilet, composting, 167
Bead board, 55, 59
Bedrooms, 175
Bermed construction
  advantages, 37–39
  disadvantages, 42
  Prairie Cocoon, 50
Bond beam. *See also* specific designs.
  definition of, 51
Bonding agents, 53
Brooder houses, 185
Building codes, 27–29, 30
Building permits, 27
Bunkers
  commercial applications of, 183
Cement blocks
  Daylight Dome, 109, 110
Checklist
  builder's, 201
  preplanning, 201
Chinese courtyard dwellings, 11, 13
Clerestory windows, 68, 105
Commercial applications, 180–194
  agricultural, 183–186
  bunkers, 183
  Freedom Tunnel, 121, 155
  manufacturing plant, 180–181
  Mobile Home Earth Shelter, 102, 104, 181–182
  office building, 189–190
  quarries, 183
  root cellars, 185–186
  schools, 181
  shopping mall, 181
  underground villages, 186–187
  university, 190–194
  Wondrous Warren, 155, 157
Concrete structures, 109–127
Conduit, 78, 79, 80, 81, 91

Convertible Crescent, 128–131
Cool pipes. *See also* soil pipes.
  definition of, 67
Cordwood Courtyard, 138–144
Cores, 51
Creosote, 85, 87
Daylight Dome, 109–115
Deep Woods A-Frame, 94-100
Domes
  Daylight Dome, 109–115
  Easy Dome, 115–119
  Freedom Tunnel, 119–123
  geodesic, 136, 157
  multi-dome complex, 119
Double Dome, 42–43, 44
Double Hex, 149–153
Drainage, 22, 42. *See also* specific building designs.
Drain tile; *see* French drain field.
Earth Shelter Digest, 9
Earth-sheltered structures
  basic modes, 37–44
Earth-shelter modes, 37–44
Earth shelters, 45. *See also* specific home designs.
Earth tunnelling, 45. *See also* Freedom Tunnel.
Easy Arch System, 145–148
Easy Dome, 115–119
Educational facilities, 20
Electricity, 18
Energy-efficient systems, 164–168
Energy savers
  heating (gas, water), 166
  wind-generating system, 164–166
Envelope construction
  advantages, 39
  disadvantages, 42–43
  semi-envelope, 40–41
Excavation equipment, 49–50
Excavation process
  Cordwood Courtyard, 139, 142
  Deep Woods A-Frame, 95
  landscape, 171
  Prairie Cocoon, 50
  Starlight Earth Lodge, 133, 135
  Trench House, 72–73, 74
Excavators, 49
Farmland
  buying, 35
  creating a park on, 36
  eroded, 34
Farrowing house, 183–185
Financing, 31–32
  funding alternatives, 36
Fire protection, 29–30

205

Flexicore
    definition of, 56
    Prairie Cocoon, 57, 58, 61, 65
    Stairway House, 159
Floor Plan
    Cordwood Courtyard, 138–139
    Daylight Dome, 114
    Deep Woods A-Frame, 98
    Double Hex, 149–150
    earth-shelter, 67
    Freedom Tunnel, 122
Flooring
    Prairie Cocoon, 65, 68–69
    Railroad Tie House, 89
    Trench House, 81
Flower beds, 170–171
Flying beam, 131
Footing. *See also* specific building designs.
    bell-shaped, 48, 55
    grid type, 42, 48
Freedom Tunnel, 119–123
French drain field, 42, 68. *See also* specific building designs.
    slope mode, 40
Funding; *see* Financing.
Fungal growth, 71
Garages, underground, 178, 179
Garden
    bermed home, 39
    courtyard, 144
    envelope home, 40
Gardening, 170
    French Intensive Method, 170–171, 172
Geodesic domes, 136, 157
Geotecture, 45, 160–163
Girders; *see* specific building designs.
Glossary, 197–203
Government land, 35
Great Plains, 84
    earth homes on, 11–13
Greenhouse
    framing of a dome, 112
    heat-storing capabilities, 176–177
    interior finish, 179
    Prairie Cocoon, 46, 47, 61–62, 63, 71
    recreation, 177
    slope mode, 40
    temperature control, 81–82, 177
Greenhouse effect, 137
Grid footing, 42, 48
Ground cover, 172
Heat gain
    greenhouse, 177
    Prairie Cocoon, 64
    Starlight Earth Lodge, 137
Heating system, back-up, 158
Heat storage, 109, 137
Indian earth lodges, 12, 132
Indigenous material, 9, 25. *See also* specific building designs.
Inspections, 27, 29–30

Insulation. *See also* specific building designs.
    cold vs. hot climates, 123–125
    in a trench, 135
Insurance, 30–31
Interiors. *See also* specific building designs.
    basic information on, 174–175
    cost-cutting, 65
Kitchen
    basic information, 174–175
    Prairie Cocoon, 66
Labor market, 23–24, 35, 94–95
Land access, 23
Land, buying
    farmland, 34, 35
    government land, 35
    quarries, rock, 33–34
    railroad sites, 34
Landscaping, 169–172
    bermed construction, 39
    envelope construction, 39
Legal considerations, 19
Light distribution
    Prairie Cocoon, 63–65
    Trench House, 78, 79
Light well
    Daylight Dome, 111–112
Living room, 176
Lumber shoring; *see* specific building designs.
Manufacturing plant, 180–181
Materials
    availability, 23, 24
    indigenous, 9, 25
    prebuying, 25, 26
    product safety, 25
    recycled, 25
    salvaged, 25–26
    shopping for, 24–25
Metric conversion table, 9
Minnesota, University of, 190–194
Missile sites, 35
    commercial application of, 183
Mobile Home Earth Shelters, 101–108
    commercial applications of, 181–182
Mortgages, 26, 31
Mother Earth News, 8, 36, 37
Mutual of Omaha Insurance Company headquarters, 28, 29
    construction information, 189–190
National Unified Building Codes, 27, 29
Nebraska, 35, 46, 47, 140, 178, 180
    underground garages, 178, 179
Nebraska State Historical Society, 13
    photographs from, 14–15
Noise levels, 173–174
    sound-muting material, 175
Office building, 189–190
Park, 35–36
Pilastre supports, 129, 131

Pioneer dugouts, 12–13, 17
 slope construction, 40
 Trench House, 72
Plants, 63. *See also* Landscaping.
 ground cover, 172
 Prairie Cocoon, 70
Plenum, 155
Plumbing, 61, 84, 174–175
Pole structures, 75, 78
Political considerations, 19
Polyethylene, 55–56. *See also* specific designs for usage.
Polyurethane, 55–56. *See also* specific designs for usage.
Post-and-beam construction; *see* specific building designs.
Posts
 placement, 73–75
 use of; *see* specific building designs.
 waterproofing, 75–76, 77
Prairie Cocoon, 45–71
 structural stress, 39
Preplanning considerations, 18–26
Privacy
 landscaping, use of, 169–170
Product safety, 25, 179
Quarries, rock
 buying, 33–34
 commercial applications of, 183
Railroad Tie House, 84–93, 95, 97, 106
Railroad ties, 84, 85, 88, 91, 103, 104, 105
Rammed-earth wall, 141, 142, 143, 144
Recreation
 considerations, 23
 greenhouse, 177
 water, use in, 177
Rectangle
 envelope mode, 41–42
 Prairie Cocoon, 45, 46
 Trench House, 72
Recycled materials, 25
Red-lining, 30
Reinforcing rods; *see* specific building designs.
Roof, construction
 Cordwood Courtyard, 143
 Daylight Dome, 111
 Prairie Cocoon, 56–57
 Starlight Earth Lodge, 134, 135–136
 Trench House, 78, 79
Roof, earth-covered, 56
Roof soil, 172
Roof, waterproofing
 Prairie Cocoon, 59
Root cellar, 83
 commercial applications of, 185–186
 tunnelled shelter, 161
Round concrete structures, 41, 109–127
 Daylight Dome, 109–115
 Easy Dome, 115–119
 Freedom Tunnel, 119–123

Salvaged material, 25–26
Schools, 28–29, 173, 181
Security, 46
 landscaping and, 170
Shear, 51
Shell fabrication, 116–117
Shineying, 171
Shopping facilities
 availability of, 22
Shopping malls, 181
Shoring
 placement, 76–77
 waterproofing, 77–78
Site selection, 18–19, 160
Skylights, 68. *See also* specific building designs.
Slope mode of construction
 advantages, 40
 disadvantages, 43–44
Slope shift, 44
Soil
 bearing capacity, 21
 composition, 20–22
 erosion, 163, 172
 percolation tests, 22
 roof, 172
 structure, 20–22
Soil, expansive
 Prairie Cocoon, 46
 Stairway House, 158
Soil pipes, 67, 68, 106, 118–119. *See also* specific building designs.
Soil pressure, 21–22
 bermed-mode house, 38
 slope-mode house, 43–44
Soil tests, 22
 for tunnelling shelter, 161
Solar chimney. *See* specific building designs.
 principle of, 68, 71
Spatial perceptions, 26, 69–70, 82, 108, 128, 173–179
Special districts, 29
Stairway House, 158–159
Starlight Earth Lodge, 132–137, 138, 139
Storage areas, 175
Sun angles, 63–64
Surface bonding
 definition of, 52
 use of, 52–53. *See also* specific building designs.
Taxes, 19
Terrace levels, 102, 105, 106, 108
Terraset Elementary School, 28–29, 173
Terratecture, 45
Thermal efficiency
 plant use in, 170
Toilet, composting, 167
Topography. *See also* specific building designs.
 earth-shelter modes, 37–44

Tractor (Caterpillar), 49, 50
Trailer park, 103
Trellis, 85, 89, 92
Trench House, 72–83, 85, 95, 97, 106
  commercial application of, 185
Trough, 51, 151
Trump Mfg. Co., 180–181, 182
Truss support system, 129–130
Tunisian underground dwellings, 11, 12
Tunnel
  entrance, 136
  Freedom, 119–123
Tunnelling, 45
  root cellar, 185–186
Twin Tees, 57–58, 60, 181
Underground dwellings
  history of, 11–17
Underground garages, 178, 179
Underground homes
  financing, 31–32
  noise levels, 173–174
  preconceptions of, 173
  selling, 26
  spatial perceptions and uses, 26, 173–179
Underground homes, types of
  Convertible Crescent, 128–131
  Cordwood Courtyard, 138–144
  Deep Woods A-Frame, 94–100
  Double Hex, 149–153
  Easy Arch System, 145–148
  Mobile Home Earth Shelters, 101–108
  Prairie Cocoon, 45–71
  Railroad Tie House, 84–93
  Round Concrete, 109–127
  Stairway House, 158–159
  Starlight Earth Lodge, 132–137
  Trench House, 72–83
  Wondrous Warren, 154–157
Underground villages, 186–187
University, 190–194
Utilities, 11, 22–23, 61

Vegetation, native
  grasses, 170
  landscaping, 172
  slope-mode home, 41
Ventilation; *see* specific building designs.
Walls. *See also* specific building designs.
  dry-stacked, 52, 53
  laid up, 51
  rammed-earth, 141, 142, 143, 144
  relationship to roof, 57
  staged, 43, 180
  tip-up, 150–152
Water, 18
  heating, 166–167
Waterproofing, 53. *See also* specific
    building designs.
  buying materials for, 25
  roof, 59
Waterproofing products, 54–55
  asphalt, cold, 53
  Bentonite, 54–55
  liquid sealers, 55, 75, 76, 93
  polyethylene, 55–56
  polyurethane, 55–56
  sheet membranes, 55
Water table
  and drainage, 22
  bermed construction, 37
  envelope construction, 39
Wind
  Mobile Home Shelter, 103
  slope-mode house, 40
  Trench House, 80
Wind chill, 39
Windows, installation of; *see* specific
    building designs.
Wiring, 61, 78, 79
Wondrous Warren, 154–157
Zoning, 20
  and building regulations, 27–28
  special districts, 29